Pre-Publication Support for "Survival through Evolution",
in order of making an overview of the book.

"A marvelous combination of hard science and of thoughts on the human condition. Very few of us have the courage in the public forum to mix convictions drawn from personal considerations with those drawn from our professional expertise. In this book, Tom Gehrels does this with rigor on both accounts, that of a scientist and that of a sensitive human being. While one may differ on some of the points of view, this book will make you think and rethink your own personal odyssey. And, if you approach it with an open mind, it may challenge, even change, your view both of the world in which we live and of yourself in that world."
-George V. Coyne S.J., Vatican Observatories

"The book invites us to celebrate the atoms within and around us whirling and twirling in vibrational patterns, orchestrating the cosmic symphony."
-Mrinalini Sarabhai, Darpana Academy of Performing Arts, Ahmedabad, India

".... a tender approach to a philosophy of peace and brotherhood."
-Fr. Danjo Souken, Saikouji Temple, Hiroshima

"This wide-ranging inspiring book on the past, present and future of human-kind in the Universe forms the essence of a much needed textbook for university audiences."   -Eric Chaisson, Director,
Wright Center for Science Education, Tufts University

"The work of Dr. Gehrels is an attempt to build on the ideas of new thinking in order to look over the horizon of current concerns toward a new ethic of global responsibility."   -President Mikhail Gorbachev, Moscow

"The history is documented of the 20,000 true pioneers of spaceflight, who gave their lives in the rocket factory of the von Braun brothers."
-Yves Béon, survivor of the concentration camp,
now living in Paris.

"I have found a respect for my fellow man in these pages that makes me want to be a better person. The book changed my life."   -Jim Wilson, student,
University of Arizona

SURVIVAL THROUGH EVOLUTION

SURVIVAL THROUGH EVOLUTION

from Multiverse to Modern Society

Tom Gehrels

BookSurge Publishing   Charleston, SC

The profits from the sale of this book will be donated by the Gehrels family to deserving charities.

Copyright © 2007 Tom Gehrels
All rights reserved.

Gehrels, Tom, 1925-
 Survival through evolution : from multiverse to modern society / Tom Gehrels.

Includes bibliographical references and index.

1. Cosmogony.   2. Evolution.   3. Social Evolution.
4. Sociology.   I. Title.

QB981.G44 2007   523.12   2007928262
ISBN 1-4196-7055-7

Visit www.booksurge.com to order additional copies.

Cover Illustration:
Protoplanetary cloud of the sort that condensed to form the solar system. This painting by the astronomer W. K. Hartmann of the Planetary Science Institute shows the accumulation of dust and rubble in the cloud. Some of the rocks still exist in the asteroid belt, in comets, and in the meteorites that fall to Earth.

Table of Contents

INTRODUCTION xi

PART I. BASIC INFORMATION 1

  1. THE WORLD AROUND US 3
     1.1 People and their Differences 3
     1.2 The Interior of the Earth and its Radiation Belts 6
     1.3 Tides and Plate Tectonics 11
     1.4 Our Atmosphere 15
  2. THE EARTH'S ENVIRONS 23
     2.1 Meteors and Meteorites 23
     2.2 Planets and Pluto 25
     2.3 The Cold and Hot Surfaces of Moon and Mercury 30
     2.4 The Hot and Cold Surfaces of Venus and Mars 33
     2.5 The Surface of Earth by Comparison 35
  3. ATMOSPHERES OF THE TERRESTRIAL PLANETS 37
     3.1 Motion and Outgassing of Molecules 37
     3.2 The Thin Atmosphere of Mars 38
     3.3 The Dense Atmosphere of Venus 40
     3.4 The Atmosphere of Earth by Comparison 41
  4. BEYOND MARS 45
     4.1 The Asteroids 45
     4.2 Jupiter and Saturn 50
     4.3 Uranus and Neptune are Different 53
     4.4 Titan, Triton and the Galilean Satellites 55
     4.5 Rings and Small Satellites 57
     4.6 Centaurs and Trans-Neptunian Objects 58
     4.7 The Oort Cloud and Activity of Comets 60
  5. THE STARS 65
     5.1 Effects of Presure and Temperature 65
     5.2 The Sun 66
     5.3 The Variety of Stars 70
     5.4 Supernovae and Black Holes 77
  6. GALAXIES AND UNIVERSES 81
     6.1 Interstellar Gas and Grains 81
     6.2 Our Milky Way Galaxy 83
     6.3 Other Galaxies and Energetic Objects 89
     6.4 Observing the Universe 92
     6.5 Expansion of Intergalactic Space 95

  6.6 The Multiverse 98

PART II. ORIGINS 105

7. THE BEGINNINGS OF OUR UNIVERSE 107
 7.1 The Mechanisms for Survival through Evolution 107
 7.2 Extreme Pressures and Temperatures 113
 7.3 The Origin of Evolution 114
 7.4 The First Minute of our Universe 117
 7.5 The Universe near Age 400,000 119

8. THE ATOMS 121
 8.1 The Miracles of the Atoms 121
 8.2 Our World is Quantized 124
 8.3 We are Hundred Percent Space 127
 8.4 Particles and Laws 128
 8.5 Waves and Matter 130
 8.6 The Cosmic Symphony 131

9. STARS AND NEBULAE 133
 9.1 Galaxy Formation 133
 9.2 Star Formation 135
 9.3 The Origin of our Atoms 138
 9.4 The Origin of our Molecules 139

10. OUR SOLAR SYSTEM 143
 10.1 The Solar Nebula 143
 10.2 Early Years 144
 10.3 Later Times 145
 10.4 The Origin of Earth and Moon 147

11. LIFE ON EARTH 149
 11.1 The Foundations for Life on Earth 149
 11.2 The Discovery of Evolution 151
 11.3 The Tools and Emergence of Life 154
 11.4 The Evolution and Variety of Organisms 159
 11.5 Further Advanced Life-Forms 164

PART III. MODERN SOCIETY 171

12. NATURAL DEATH 173
 12.1 Multiverse and Galaxies 173
 12.2 The Solar Neighborhood 175
 12.3 Death of the Sun 175
 12.4 Hazards due to Comets and Asteroids 177
 12.5 Death and Rejuvenation 182

13. MAN-MADE PROBLEMS AND SOLUTIONS ... 185
13.1 Nuclear War and Radiation Effects ... 185
13.2 Chemical and Biological Warfare ... 189
13.3 President Eisenhower's Warning ... 190
13.4 The Global Warming Emergency ... 197
13.5 Over-Population and its easy Remedy ... 204

14. TECHNIQUES FOR SOLUTIONS ... 207
14.1 How to diminish Cruelty ... 207
14.2 Healthier and more cheerful Living ... 210
14.3 People find Solutions and make Resources ... 215
14.4 The Human Genome and Genetic Engineering ... 217
14.5 Molecular Nanotechnology ... 220
14.6 Space Exploration, after gruesome pioneering ... 222
14.7 Human Space Travel ... 225

15. ADVANCED LIFE ELSEWHERE ... 229
15.1 Searches for Planets of Other Stars ... 229
15.2 Are we Alone in the Multiverse? ... 231
15.3 How Long will an Intelligent Society Survive? ... 232
15.4 Unidentified Flying Objects.(UFOs) ... 234
15.5 Searches for Extraterrestrial Intelligence ... 235
15.6 How would we Communicate? ... 237

16. WHAT IS NEXT FOR PLANET EARTH'S SOCIETY? ... 243
16.1 Coming Together ... 243
16.2 The Rise and Decline of Societies ... 244
16.3 Survival of the Thoughtful ... 246
16.4 The Evolution Guideline for Action ... 247
16.5 Integrity among Nations ... 248
16.6 Learning from Opponents ... 250
16.7 New Solutions for Poverty ... 254
16.8 We can Win ... 257

REFERENCES AND NOTES ... 261
ACKNOWLEDGMENTS OF PEOPLE ... 284
ACKNOWLEDGMENTS FOR ILLUSTRATIONS ... 285
GLOSSARY ... 287
INDEX ... 307

adagio molto e cantabile

Introduction

    Beethoven's Ninth Symphony masterpiece starts confidently, but then, before making its ultimate triumph, it slows down very carefully, 'molto adagio', and yet with a melody, 'e cantabile'. It seems to be Beethoven's Prayer, at least one outstanding musician agreed, and so it became the theme for this book, which is to be an interlude in our busy lives for contemplation of our world and ourselves in it, our origins and the future with its many hazards. The cantabile is for the beauty of nature, even though there also are ugly aspects in our world. Can we overcome them? Yes, after our dedication with a melody of information, we will end as Beethoven dared, with an ode to joy.
    The technique we will use is in the title of this book, the combination of survival and evolution. Evolution helps our survival, and the opposite is also true, that survival is the basic concept of evolution by bringing a way out towards the future. Ask any mother about the future when she holds her newly born.
    The word "evolution" came long ago from the Latin 'evolutio', which means 'unfurling,' whereby our lives unfurl in a progression of interrelated events. 'Multiverse' in the title is a rather new word, meaning all universes, ours included. After some description we will see in this book how the evolution of the multiverse decided the characteristics of our universe and thereby of our existence and of the evolution of modern society.
    One cannot think or speak about evolution without education in the topic, as is true for any topic or trade. So, here is a handbook like every trade has one. It has a light touch with a serious line through it of reality. In any exchange of ideas, as in evolution, it is important to have an understanding of the words that are used. The people who are debating must be clear about the topic and be sure that their debate is about the very same concepts. Therefore, any key word we use here will be found in a dictionary, and we will often refer to the Latin (L), Greek (Gk) or other roots of a word, as we did for 'evolutio'.
    This book is about observations of nature, from a beginning before

our universe was born, to the origins of what came afterwards, and then to global problems of human societies. We are living in a violent world. Weapons are everywhere and new types are being designed. Poverty is also everywhere, awesome global problems and deadly divisions are lurking, while most people and their leaders have no common background or guideline for making improvements or for finding solutions. I propose here to try to develop such a background from simple observations. Because of the central position of science in our civilization, scientists should communicate their views and become involved in cultural and political affairs[1]. Their outreach to the public is encouraged by universities and by the agencies that support their research.

It is high time for humanity to formulate its guidelines from an appreciation of its place on this planet; a cosmopolitan worldview is required because our problems are spreading worldwide and becoming truly dangerous. The world and its future need a shared approach to global problems, by integrating them with what we know of nature and its evolution. When all three are studied together - nature, people, and their problems - it is seen that they are thoroughly interactive, and that this therefore becomes a good approach towards making a common background and guideline. These sentiments are in the banner of the Web site of the former President of the Soviet Union, Mikhail Sergeyevich Gorbachev, based on his global experience,

> We need a new system of values
> a system of the organic unity,
> between mankind and nature
> and the ethic of global responsibility.

The University of Arizona has provided a beginning with that pursuit in an undergraduate course for non-science majors, and I schedule my sessions in the evenings so that teachers and other people with daytime jobs can participate in the discussions. Expansion of the outreach occurs through my involvement in India, which includes a United Nations' Course for graduate students from countries as far away as Kazakhstan and North Korea[2].

We will prepare ourselves by over-viewing our world in Part I of this volume, viewing globally and then ever farther out into our solar system, galaxies and the multiverse. People interact politically and economically with nearly everyone else on Earth, but many lack a knowledge of planets, asteroids, stars and nebulae, which is important for understanding our birth and abode and how fragile they are. We will concentrate in this book on features of our global environment that are important for understanding humanity's place in the universe. The observations of Part I serve also to become familiar with the terminology for our origins, which are covered in Part II, from the largest scale of universes to life forms. Finally, in Part III, we use these perspectives to study our global problems and to suggest some guidelines for possible solutions and resources.

You will become impressed with the intricacy of nature, but also with the striking feature that basic components are used over-and-over again, like bricks for a building and letters in the DNA, such that the information becomes easy to learn and to remember for life. The beauty of nature is ethereal, how we are not composed of inert atoms as the Greeks believed, and many people still do. People might believe for instance that a mother's womb fabricates the atoms for a child - but we cannot produce a single atom, for they are far too powerful. I will not give you much math in this book, and the numbers are rounded off so they are remembered for life. However, it is essential to know that we consist of atoms, molecules and cells made of quantum-mechanical space-waves that are interacting with each other through their wave structure. It will become natural to view yourself in such an interactive world. The quantum stuff is not difficult to learn. I will lead you through an introduction such that it will be easy to take it all in for pure enjoyment, like reading a novel wondering how it is going to come together.

Our future has some frightening hazards so that we search for at least the origins and principles of these problems and their solutions. The same methods are used in Part III, giving observations and descriptions so that you can decide the various answers for yourself, and when there is a debate or an argument, both sides will be presented. For example, we will have a great time in Chapter 15 finding optimists and pessimists regarding the question of whether we are alone in the multiverse. However, much of the two opposing views come from peoples' feelings more than from the few facts we have on this topic.

Our insights are based on summaries of observations in nature, i. e., on **natural laws**[3]. It may be the first time in history that such a worldview is possible, because the coverage of the sciences has now become broad enough at the scientific level of this book. Imagine the amount of work that has gone into this worldview. Millions of scientific and technical papers provide the basic background for this book. However, our view cannot be complete, because each era can penetrate only to its own depth of perception. One should not pose questions that cannot be answered scientifically as yet, like "how did the multiverse originate?" We are actually glad that this is so because then there are always attractive fields and questions for new exploration.

Anyway, those natural laws yield the possibility of having an internationally shared view, because the observations of nature are nearly the same by careful observers anywhere, be they in Tucson or Ahmedabad, Cairo or Kyoto, or wherever you are and whatever your background may be. Variation in background and individuality of the observers may influence the interpretations; a simple example of that is the insistence on Pluto being a planet, or not, for which we will again show the emotionally opposite sides. What helps here is the reviewing and refereeing of major conclusions before publication in respected journals. The result is a near-consensus for the broad summaries that suffice at the level of this book. The new knowledge and understanding appear promptly and everywhere in electronic form, such that a modern worldview becomes a new generation's birthright.

Many people have helped to improve this manuscript and they are in the Acknowledgements Section - I thank you all. The Glossary has some definitions, and the Notes give references for further study; a few Web sites are indicated, trusting that the reader can take it from there by surfing the Web[4]. In the text, the first sentence of each paragraph will usually summarize it, and the main topics stand out in **bold face** within the paragraph.

It would also be appreciated if experts who are already familiar with the topics in this book would check the observations and interpretations; such refereeing is essential when the ideas are new. On the one hand, this book is mostly based on well-established science. On the other hand, I do present my own views, but I specify here where they occur, namely in Sections 1.1, 6.6, 7.1, 7.3, 8.3, 8.5, 8.6, 9.4, 12.1, 12.4, 12.5, 13.3, 13.5, 14.1, 14.2, 14.6, 15.4, 16.3, 16.4, and 16.6 - 16.8. Please send corrections, disagreements, and information missing, so that a revised edition of the book can be made soon.

Making this text was a thrill of discovery and I want you to experience that too. Join me then for a guided tour through space and time.

Tom Gehrels

Part I. BASIC INFORMATION

1. THE WORLD AROUND US

1.1 People and their Differences

Let us begin with an example in which an understanding of nature and humanity is helpful for doing anything about segregation problems in the world. Much is made, and strife results from the fact that people appear to be so dissimilar, having darker or lighter skins, adhering to different religions, and speaking an endless variety of tongues and accents. However, these differences have been brought about through **circumstance**. The roots of humanity developed near the warm **Tethys Sea** with its exceptionally good conditions for the evolution of life; that is where the oil fields of the Middle East are now (Fig. 11.1, below). As the result of plate tectonics, we see landmasses sliding over the early bio-growth, so that the weight and pressures decomposed the bio-mass into oil and gas (compare with Fig.1.5, below).

More advanced life began to develop fast some ten million years ago, particularly on the Western shores where the East African Rift Valley is now, near the present Kenya, where abundant rainfall and growth of food and offspring had occurred in early times. The area lies in a domain of active plate tectonics that first had brought mountain ridges and their rainfall ("orographic" rainfall caused by the air rising up the mountain). However, the direction of plate motion slowly reversed and began spreading apart, the climate became drier and the people began to search for places to live elsewhere, migrating from their birthing grounds to a wide variety of climates, where they usually settled in isolation[1].

The migrations that happened a few million years ago eventually reached all over the globe. However, where they found good places to settle, they did settle, and there they stayed usually in **isolation** for millennia. There we **see** the cause of different mother tongues, skin colors, and religions. Isolation of the various people on Earth is the key. Solar radiation affects skins strongly, while foods and fears come from the environment, so that, over time, various populations of the world became different in appearance and behavior.

I have seen extreme cases of isolation in 1946 for the Papuas of New Guinea, and still in 1972 for the Tara Umara tribe in Northern Mexico. The Papuas were rarely communicating even with people in the next valley, let alone dealing and trading with foreign tribes of distant islands and territories. The ones we visited would stay up all night near the time of full moon to beat their drums, because falling asleep would show disrespect to their God, who would punish them with awful accidents in the following month. **The moon is their god** and that made us Europeans recall our own religious origins when the sun was one of our gods. The name of our Sunday had originated that way together with names for weekdays in honor of the other gods worshipped by our fore fathers. The Tara Umara would have single families in open and windy caves high in the mountains, such that their life expectancy was only 35 years.

It is easy to remember the grouping of the main **physical differences** between people. The world's population divides in terms of appearance approximately as follows: Sixty percent have brown skin and black hair and live mostly in Asia. Twenty percent have light skin and "Caucasian" is an indication for them, probably because they originated not far from the Caucasus Mountains and then spread in all directions, westward all the way to the Atlantic Ocean; the light skin emerged only some 10,000 years ago. Ten percent reside in Africa, but also New Guinea, and they have black skin and curly black hair, presumably caused by extreme solar radiation. The remaining ten percent living in middle and South America have mostly brown skin and black hair.

The distribution of adherence to a religion or **belief system** is simpler to remember since the world's population happens to be distinguished in that respect roughly into **quarters**. What follows is only for easy remembering of an interesting conclusion, with due apology to anthropologists as the Islamic and Judeo-Christian religions emerged from the same roots and have several prophets in common, and that Hinduism and Buddhism are greatly different. With that respectfully in mind, one quarter is seen for the part of the world where Hindu and Buddhist religions are followed, another quarter follows the Quran, while the third quarter bases its beliefs on Judeo-Christian texts. The remaining quarter belongs to a variety of other religions or to none at all. Religious refinements and separations have increased greatly over the millennia. Taking the many subdivisions of the religions into account, it follows that the beliefs of any one person today, like you or me, are shared by less than **10%** of the world's population. Not more than about one-tenth of the world's population believes what any one of us believes.

However, we should always remember that underneath these **acquired variations** lies the commonality of our atomic material, and that is the same for all of us, 100% the same. Furthermore, we are **not isolated anymore** thanks to modern travel and electronic communication, such that we have an opportunity to understand and ameliorate the differences.

The stark plight of the very poor is of the greatest concern in our consideration of the world's future. Table 1.1 lists my estimates of the equivalent in U.S. dollars that is available per day for various numbers of individuals on Earth. I took into account that the official exchange rate for foreign currency to dollars is appropriate for the needs and productivity of people who use modern apparatus and conveyances. For the destitute, however, most of them living in tropical climates, the needs are simpler and taken care of locally. They may obtain some food for say 4 Indian rupees, which is only 10 dollar-cents on the exchange rate, but if we would buy something in modern supplies and conveniences, as for a part of a bike, that would indeed cost a dollar, so I wrote $1. I call this the **exchange effect**. If one would contribute $1 in poverty relief to a local organization like Akanksha (Sec. 16.7) that has little overhead costs, the gain would be a factor of 10.

Table 1.1. The Distribution of Incomes.

Number of Individuals	Estimated Daily Income
1×10^9	$1
2×10^9	$6
2×10^9	$15
1×10^9	$60
1×10^7	>$600

The second line in the Table is still one of poverty, even though these people may use a bus or have a bike. The exchange rates may therefore begin to apply in part, and they certainly apply in full for the third line. I obtained the amount for the fourth line in Table 1.1 by guessing a yearly income of about $65,000 per year, per average family of three (having 3 children that are, however, dependent only during one-third of their parents' lifetime). The bottom line is simply my guess; > means "more than".

We should note that there are wealthy people even in Third World countries and that there are homeless in the Western world as well. The ratio of highest to lowest income has some surprises as is seen in Table 1.2.

Table 1.2. Ratio of the income of the richest 20% of population to that of the poorest 20%, 1981-92 (for the U.S. in 1993)[2].

Country	Ratio
Brazil	32
Guatemala	30
Senegal	17
Mexico	14
United States	13
Malaysia	12
Zambia	9
Algeria	7
China	7
South Korea	6
Germany	6
India	5
Japan	4

The total number of people on this planet reached the even number of **6 billion** in November 1999, 6.0×10^9. The number had been doubling every 40 years or so, but the rate of increase is now dropping steeply as birth control is beginning to take hold nearly everywhere; we shall discuss this remarkable result in Sec. 13.5.

Incidentally, the word 'billion' is used and understood only in North America. For Europeans, it is a confusing word because they have a similar word for a thousand-billion, so we will occasionally remind ourselves that a "billion" is a thousand **m**illion. The spacecraft Mars Climate Orbiter crashed soon after launch because its engineers had failed to understand international conventions. This should be an expensive lesson that the world has become too small for such lack of education. As far as large numbers are concerned it is actually easier and more certain to use **exponential or 'scientific' notation**, for instance 10^9 for the American billion, and 10^{12} for the above European "billion." An overview of exponents is in Note 3 of this Chapter[3]. Now we will begin our tour through space, near or inside the earth at first.

1.2 The Interior of the Earth and its Radiation Belts

We begin with several observations showing that the earth is not a rigid body[4]. We might have expected a perfect sphere because gravity pulls its mass uniformly together from all directions towards the center of gravity. However, there is an additional effect caused by Earth's 24-hour rotation and this causes a slightly larger radius at the equator than at the poles. It is a small difference, only 0.3 %, but it is important to understand as a fleeing from the center, a **centrifugal** effect, like the one that we see and feel when swinging a weight at the end of a rope around. On the earth, the direction of the effect is away from the axis of rotation, and it is therefore absent at the poles, with the result of a somewhat flattened shape pf the earth rather than that of a perfect sphere. The effect has been known for a long time, but the explanation has become clear more recently, that the deformation can happen only because the earth is not rigidly solid. The deformation is possible because of high temperatures deep inside, so much so that the interior must be **molten,** at least in part.

When you have a chance to go down into a deep mineshaft, you will indeed notice that it gets hotter rapidly. As a rule of thumb, the temperature of the earth increases one **degree Kelvin for every kilometer** closer to its center. Since the earth's radius is approximately 6000 kilometers, the temperature near the earth's center should be 6000 K, and it is. A refresher on density and temperature is in Note 5 of this Chapter[5].

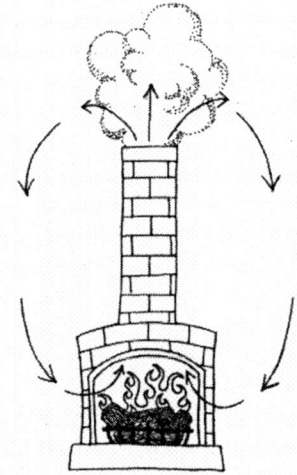

Fig. 1.1. The principle of convection.

At 6000 K, the material must be molten or at least we expect that it can move around like a gel, even at the great pressures of the weight of all the layers above. In fact, the internal heat drives some of the material into slow motion outward towards the surface of the earth where it is cool. Near the surface, the material will then cool down somewhat; the molecules will become less agitated and collide less frequently. After more cooling, the slowing of the molecular action makes it all become more tightly compacted, denser and heavier so that it begins to settle downward towards the center again. Such a looping process we see in a chimney in Fig. 1.1, while Fig. 1.2 is an illustration of what happens inside the earth; **convection** causes the looping, an important concept we will encounter again in the atmospheres of planets and stars. The internal heat causes the rising – the cooling above causes the material to become denser and heavier in order to descend again.

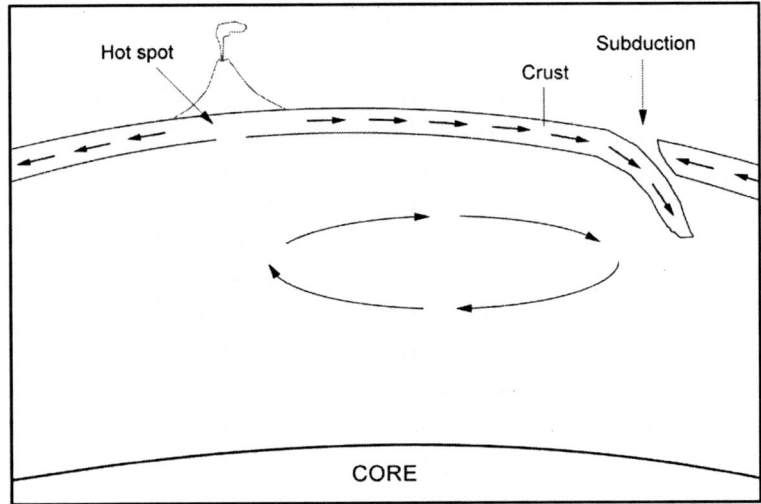

Fig. 1.2. A sketch for the concept of convection in the earth's mantle.

The pressure and agitation at the highest temperatures near the center make the molecules **collide** so violently that their molecules suffer some damage, and some of the outer electrons break free from their molecular bonds. Now we have the miracle inside our Earth, of free negative electrons and of positive charges too because the molecules have lost some electrons, and we speak of an electrically charged medium. Next, we remember the natural pair: **motion of electric charges generates magnetic fields,** and *vice versa*: **motion in magnetic fields generates electricity.** We use such **electromagnetic** effects in our daily lives in many applications, for instance in electric motors and in the electricity generators of automobiles. In the case of the earth, the motion of the electric charges produces the magnetic fields that we see and use with a compass.

The magnetic field of the earth is not steady and constant however, because of random variations in the convection patterns in the interior. At present, the giant terrestrial magnet happens to be approximately in a north-south direction, so that a compass points towards the magnetic south and north poles. These poles do not coincide exactly with those of the earth's axis of rotation. The **magnetic North pole** lies near Greenland; the magnetic South Pole, south of Tasmania. The earth looks and behaves like a huge magnet (Fig. 1.3)[6].

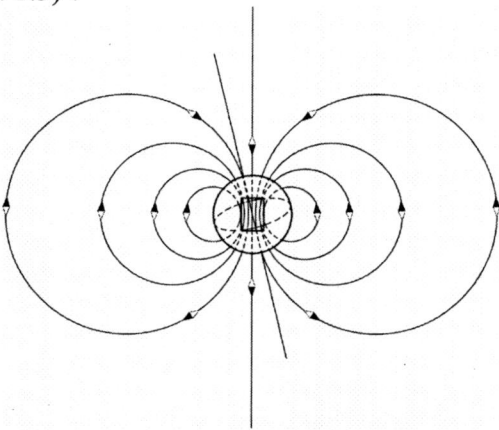

Fig. 1.3. The earth's magnetic field.

The magnetic fields extend well beyond the surface of the earth, as is seen in Fig. 1.3, and there they interact with and protect us from the electrically **charged particles** that come from the sun. The sun is a strong source of protons and electrons[6]; we speak of the **solar wind** while at times the sun flares out to produce super violent **solar storms** (see Fig. 5.1, below). Their study and of their effects on Earth is an active field of research, even in 3 dimensions by two spacecraft in the program that is called STEREO, for Solar TErrestrial RElations **O**bservatory[6]. When the protons and electrons approach Earth, even at speeds near a million kilometers per hour, two donut-shaped belts halt and deflect them (Fig. 1.4). The cause of their deflection is that they also are electrically charged; as you know, the protons have a positive charge, the electrons negative. They thereby interact with the magnetic effects, spiraling around the field lines of Fig. 1.3, which propel them towards the magnetic north and south poles. Through this process, the magnetic field accelerates them rapidly so that they become dangerous high-energy particles again, but now kept high above the humans who live safely below. The two donut-shaped radiation belts are in the Glossary under **Van Allen Belts,** named after their discoverer, Physicist James Van Allen of the University of Iowa, who was using the sounding rockets of Sec. 14.5 that were carrying his own instruments.

Understanding the belts is important because of their role in protecting life on Earth. The astronauts, who ventured beyond the belts to the moon in the 1960s and 70s, were fortunate because there was no solar storm during their journeys, but for longer trips like to Mars one cannot predict that. The **astronauts' health** - and that of their offspring if they had produced any after their flights – might have been seriously hurt by high doses of irradiation.

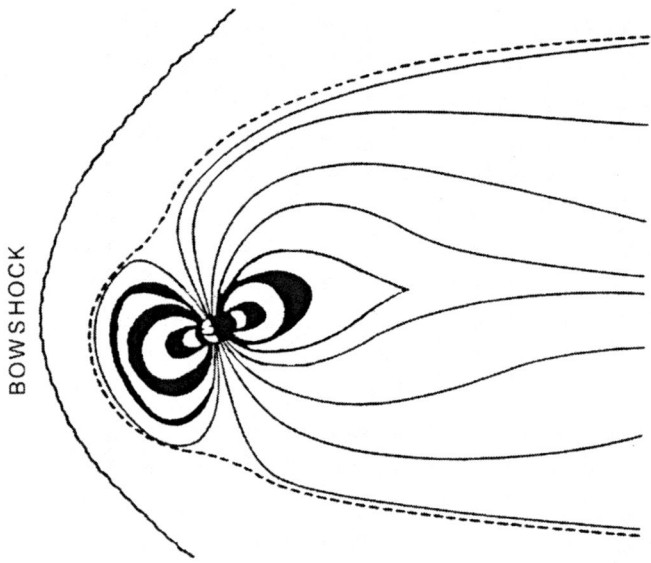

Fig. 1.4. The Van Allen belts.

Our Figs. 1.3 and 1.4 also show how the field lines veer together and concentrate near the magnetic poles, so you can imagine how densely crowded the charged particles become because of the convergence from all directions. It shows that concentration is essential for the mechanism of the belts, because there will have to be enough of them, crowded enough, to reflect and bounce the charged particles back towards the other pole. During a solar storm, however, the charged particles may suddenly become so densely pinched together that they leak through, down into the earth's atmosphere. And then their impacts on and electric interaction with the atmospheric molecules result in a break up, and recombination too, of the molecules, and these interactions cause glows of our earth's **northern** or **southern lights**, also known as **aurora borealis** or **aurora australis**. When you fly in a plane at night near a magnetic pole, as commercial jets do from

North America to Europe, make sure you have a seat by a window on the *north* side (pilots stay well south of the Canadian isolated wilderness). Carefully shield your eyes from cabin lights, and you may be rewarded with an unforgettable display of huge curtains of light varying their shape, color and brightness with glorious curving and waving patterns and all of this happens on time scales of merely minutes[7]. Let us come down to the earth's surface and inside it again.

1.3 Tides and Plate Tectonics

The levels of our seas and oceans are not constant, and that is due to the combined gravitational attractions of the moon and sun causing the **tides**. The moon is much less massive than the sun but is much closer so that its tidal effects are about twice as strong. The times of the tides therefore shift primarily with the motion of the moon, and because of its orbit around the earth we see motion eastward of both the moon in the sky and the tides in the oceans, coming about 53 minutes later each day. But there also are the times near full and new moons when the **combined** influence of sun and moon produces unusually high and low levels called **spring tides** (not like in the spring fling, but for action like of a spring). The difference in water levels between high and low tides is about one meter (40 inches) for the oceans on average, but it varies because the levels are locally affected by the depth of the sea and the shape of bays, fjords and inlets. An exceptionally large difference of 8 meters between high and low tides happens near the estuary of the Colorado River, while at places farther into that Gulf of California, the "Sea of Cortés," there is almost no tidal effect at all[8].

Note that there is also a tidal effect on the **surface** of the earth, which moves at least 25 cm between low and high tides. Yes, on our "solid" Earth we are **moving up and down at least a quarter of a meter daily**, demonstrating again that planet Earth is not a rigid rock. That is an impressive observation, readily made with a small instrument.

At this point on our guided tour, we must make a quick trip backwards in time through thousands of millions of years in order to understand the layering of the earth and to appreciate the fossils in those layers. Over such long times, the rains gradually wash silt and pebbles from mountains and plains down the river and into the seas and oceans - this is called **erosion**. Well inside the ocean where the rivers cease to flow, the **sediment** deposits in successive layering, with the more recent arrivals above the older ones of course. Next, the convection in the mantle of the earth that we explained with Fig. 1.2 may lift that sea floor up. A river that keeps running when such an uplift occurs will be cutting deeper and deeper into the deposited layers; such was the case for instance for the Colorado River cutting away the depths of the Grand Canyon when there was an inland sea draining through that area. The visible sides of the river will then reveal millions of years of layered history; such exposure occurs in the same manner where roadways cut through a mountain. Some of these layers will have solidified into rock or compacted soil, thereby preserving the entombed remnants of organisms that may be extinct, the **fossils**. Along such riverbeds or highway cuttings the layering of the ancient deposits is clearly visible and the fossils can be collected for studies of origins and evolution.

Let us next consider some consequences of the fact that molten substances occur in the interior of the earth. We saw evidence of this in the earth's shape and in the tides' vertical motions of our "solid" surface. The molasses-like material allows the heavier substances to seep through and settle deeper down towards the center through the force of gravity. In a simple fun example we see the effect even over small distances in a glass of coffee after adding milk if left unstirred - the milk lies below the lighter coffee water. It is amazing to try this remarkable experiment and show it to others in conversation. **Differentiation** is the name for this effect, and it is a logical word because it makes gravitational differences in layering. In the earth, the heaviest substances are the metals and they have sunk and settled towards the deepest regions of the interior. The density of terrestrial material therefore increases from that of lighter and looser sandy material on the surface to the heaviest most compacted metallic material near the center, where it produces the magnetism we described earlier.

During the earth's early stages, the molten surface cooled first on the outside of course, forming continental crusts and ocean floors. Parts of the crust and floors may drift over the deeper partially molten layers so that we speak of **plate tectonics** -- from the geological term "plate," meaning a large area of the solid crust of a planet, and the Greek word "tekton," which means "builder." "Plate tectonics" is preferred over the older and familiar term "continental drift" because the drifting takes place for ocean floors as well as for the continents, as is shown in Fig. 1.5.

We see that the "solid" ground beneath us is actually in motion, vertically as well as horizontally, and we understand that the heaving and shifting of tectonic plates affects our environment[9]. Reference 9 is for beautiful National-Geographic diagrams of plate tectonics. Figure 1.5 merely shows a summary, but that you can always remember of how the plates move and push against each other. Some of their distant past is shown in Fig. 11.1; comparison with Fig. 1.5 gives a feel for the movement of the plates. Please study Fig. 1.5 closely because the various parts of the world from which we hear of volcanoes and earthquakes show an enormous amount of detail regarding the causes[9]. The rate of motion of the plates ranges near **5 centimeters per year**, which is about twice as fast as the rate of growth of our fingernails. The arrows in Fig. 1.5 show the directions of motion, and they give us a feel for the variation in convection patterns underneath the plates, inside the mantle. It can now also be understood from Figs. 1.5 and 11.1 that the magnetic fields of the earth are not steady because the convections change in restless motion, albeit it slowly with time. The following paragraphs will discuss five effects that are representative for plate tectonics.

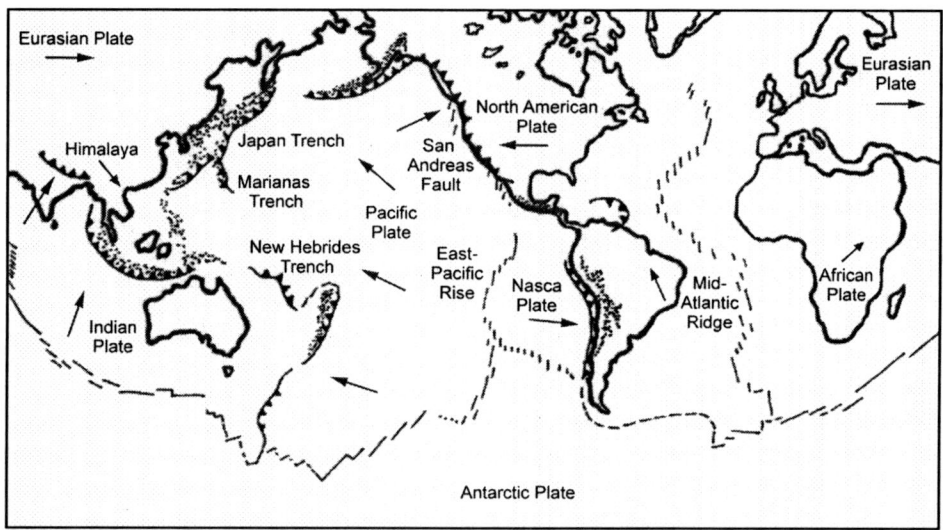

Fig. 1.5. Sketch of the motions of tectonic plates.
This Figure is worth detailed study of volcanism,
earthquakes, and ocean rifts near boundaries of plates.

The Indian Plate is pushing northward against the neighboring Eurasian Plate, which does not share that motion at all, but goes eastward. No ocean is involved any longer, and the collision of the two plates lying at about the same level began 50 million years ago, as has been determined recently[10]. The Indian Plate seems to have been shoving into the Eurasian to an enormous extent, clearly seen in the great elevation and size of the Tibetan plateau[9]; a good atlas shows the effects as far as Thailand and

Myanmar[9]. The folding and wrinkling of the colliding parts of the earth's crust later wrought the Himalaya mountains, which began to rise as the result of the tectonic forces, but only about 10 million years ago, apparently happening only after the Tibetan action was saturated[10]. The Himalaya are continuing their rise, but their top soil erodes away, is carried down by the rivers, and this process just about annuls the gain. In general, **mountain building** occurs at various places on Earth; we referred to the effect it had on early human evolution and migration at the very beginning of this Chapter.

The North-American Plate is moving mostly due **west**ward whereas the neighboring Pacific Plate drifts in a **north**westerly direction. The contrary motions clash at the "San Andreas Fault", which runs through California from San Francisco to east of Los Angeles, where there is stress and friction between the plates. Occasionally there is a sudden release of stress and that causes **earthquakes.** Geologists have predicted a major earthquake to occur in California in the near future because there has been no significant release of fault-line stress for some time.

The measure of an earthquake's magnitude is on a logarithmic scale, for instance "the Richter Scale" after Charles R. Richter (1900 - 1998), the seismologist who invented it. A refinement of Richter's scale, **the moment scale,** is presently in use for expressing the strength of earthquakes[11]. The smallest reported is 3, the largest 8, or 9 in exceptional cases, and we will see one below.

The lines near the middle of the oceans in Fig. 1.5 show **oceanic ridges** where plates are separating and drifting apart so that flows of gas, liquid and magma from down below appear, and cool in the ocean water, and the magma then solidifies around these **deep-sea vents**[12]. The crust at the sea floor is thinner than the continental crust: oceanic crust is uniformly near 8 km thick, continental crust averages 35 km but is more variable, reaching 60 km under the Andes and Himalaya Mountains.

Tectonics effects also cause **volcanic activity.** The convection below a plate may shove a plate underneath another one if it is already lying low, being at an ocean bottom. Then it moves still deeper downwards, towards higher temperatures so that water evaporates and soil and rock starts melting. The phenomenon is called **subduction** ("sub" for "under" like for a submarine; and "duction", as in "conduction", suggesting movement). The heated matter expands because its molecules move and collide more frequently and with greater energy, so that the expanding magma of steam and lava will find or force a passage upwards and burst forth as a volcano. There is a small ocean-bottom plate subducting Eastward underneath the states of Oregon and Washington where a North-South line of volcanoes is seen. The melting of the plate going eastward seems to occur at a certain depth of about 150 km, a remarkable effect so that we see the line of volcanoes running parallel to the Pacific Coast, **the Cascades Mountains**. Mount Saint Helens blew its top in 1980 and became active again in 2004, while Mount Rainier near Seattle is expanding at times, the whole mountain, and one expects an eruption to come soon.

Subduction effects are much stronger at the west side of the Pacific as you would expect from the opposite motion arrows of Pacific and Eurasian plates. Volcanoes and violent earthquakes occur in Japan and especially in Kamchatka; there are deep **trenches** near the western shores of the Pacific Ocean showing the effects of the subducting plate going down. Only subduction seems to be able to produce earthquakes with moment-scale values as high as 9, and this is what happened in the eastern Indian Ocean in 2004 where the Indian Plate is a low ocean bottom, sliding underneath the Andaman and Nicobar Islands. Then, suddenly, its friction slipped free, causing the 9-scale earthquake. The sea bottom near the islands had moved up, heaving one of the deadliest **tsunamis** (from the Japanese tsu meaning harbor and nami meaning wave) of the water felt all around and beyond the Indian Ocean.

The Hawaiian chain of islands in the Pacific Ocean is emerging by the drifting of the Pacific Plate in northwestwardly direction over a **hot spot**, deep in the earth's mantle (Fig. 1.2). Hot spots are probably caused by occurring over a **plume** of exceptionally hot material that rises from deep in the mantle, perhaps even from the core-mantle boundary (more about hot spots is in the Glossary). The Hawaiian chain of volcanic islands emerged as the plate moved over this plume, Midway Island first, and the latest, eastern-most, the big island of Hawaii. Visitors may sometimes stand close to a vent on the very eastern side of the island where red-hot lava comes pouring from the surface and cooling at some distance; it is a stirring experience actually seeing Earth in the Making. The next island-in-the-making, the Loihi Seamount, has an elongated shape and "long" is "lo" in Hawaiian. It lies 28 km towards the southeast from Hawaii's Kilauea Crater. Loihi's summit presently reaches less than a kilometer below sea level; if it remains as active as Kilauea is today it will emerge in a few tens of thousands of years[13]. Now, let us look upwards, at the air above us.

1.4 Our Atmosphere

For the terrestrial atmosphere, let us first use what we know from flying as passengers in commercial aircraft. Half of the mass of the earth's atmosphere lies in its lower 5.6 km, while modern passenger planes cruise well above that, at about 8.5 km. Aircraft altitudes are reported and prescribed all over the world in **feet**, peculiarly enough, and the conversion is 5.6×10^5 cm / 2.54 cm per inch / 12 inches per foot = 18,400 feet (a **kilo**meter has 10^5 **centi**meters, there are 2.54 cm per inch and 12 inches per foot). East-west flights are at whole thousands, like 38,000, and north-south flights in between, like 37,500 or 38,500 feet, so there is 500 feet separation for safety.

The practical limit of the atmosphere lies near 100 km (328,000 feet), and you can tell (from the above fact that half of it lies below 5.6 km rather than at 50 km) that there is not a simply straight linear relationship. Instead, atmospheric pressure decreases steeply with altitude, not linearly but exponentially. This remarkable situation is due to the simple fact that **a gas is compressible**, unlike soil and water. The lower layers of our atmosphere are therefore dense enough for breathing happily, while higher layers quickly become rarefied to the point of being deadly.

Such low atmospheric pressures seemed a severe problem for high manned flight in its early days. When test pilots first flew in jet planes at high altitudes, they needed oxygen masks already at 10,000 feet, and they would have died at altitudes greater than 40,000 feet without a pressure suit. Now however, they, and all passengers, fly in pressurized airplanes comfortably without perhaps even being aware of the miracle of their safety. A deadly environment occurs right outside their little window, an extremely thin atmosphere at temperatures below -40 degrees (the point where Celsius and Fahrenheit happen to be the same). Should their cabin spring a leak at that altitude, breathing through oxygen masks is the only way to remain conscious, or people will **pass out quickly**. This is why the flight attendant tells the passengers to put their own oxygen mask on first, and only then take care of the children. I wonder why the passengers are not being prepared for the other miracle, of what will happen at that time to the airplane, a steep but controlled descent to safer levels, and the pilots rehearse that regularly in flight simulators. When you arrange for your flight as a passenger, always ask for a window seat **on the side where the sun is not**, for there is so much to see even during normal flight[7].

The composition of the earth's atmosphere is approximately **three-quarters nitrogen** and **one-quarter oxygen**. Oxygen is active and effective in our breathing and intake into the bloodstream. Nitrogen is less aggressive; it is so abundant by default because it is too inactive to be absorbed much into the rocks and oceans. Small amounts of other molecules are also present in the atmosphere, and, of course, water vapor to make clouds which vary with place and time, as we know so well.

Another molecule of special importance is **ozone**, which is concentrated in a high-altitude layer near 30 km (98,000 feet). The ultraviolet radiation from the sun breaks up the oxygen molecule, O_2, into two oxygen atoms, which combine with another free oxygen atom to form ozone, O_3. **Ultraviolet** solar energy is thereby absorbed into the ozone of the ozone layer so as not to reach the surface of the earth. Our eyes evolved in this sheltered environment with the result that they cannot detect ultraviolet light for there wasn't any. Ultraviolet light has short wavelength, it is energetic, aggressive, and the eyes did not evolve a protection or defense mechanism for that because there was no need for that. So now, the welders at work must shield their eyes with a mask designed to absorb the short-wavelength ultraviolet radiation that comes from their highly energetic welding. We, however, without such protection, must not look at that welding glare. Should the ozone eventually disappear because of our pollution, as it will soon if we do not stop that, we will be in big trouble.

The sensitivity of the eyes of people evolved over the color range from the violet (but not ultra-violet) through blue and yellow as far as deep red (but not infra-red), and this range is therefore called the **visible spectrum**. This result was efficient, brought about by adaptive evolution to the largest amount of sunlight, which occurs over that range of wavelengths. The efficiency is beneficial especially at the hazardous times of twilight, and the eye even developed components of the retina for seeing dark and bright objects the best, which are called the **rods** and the **cones**, respectively. However, it is too bad that we cannot see what the world looks like at other wavelengths, as some animals can at colors a little different. The visible spectrum is small, narrow, in comparison to the widely ranging variety of radiation of the universe. Figure 1.6 shows the total range of energies and forms of radiation - please study it in detail for it is important for understanding the international space programs. The vertical axis in **meters** is a good unit for long-wave radio and television engineering. For atomic scales the appropriate unit has to be much smaller, **nanometers** (1 m = 10^9 nm; 1 nm = 10^{-9} m); the word "nano" is understood at a glance at the Powers of Ten Table in Note 3 for this chapter[3].

The above limit of 100 km to our atmosphere means of course also that at that level lies the limit to atmospheric absorption of the light from sun and stars, etc. for that is where the number of absorbing molecules has practically vanished. Sophisticated spacecraft and instrumentation, flying well **above 100 km,** are available from international space programs for the study of astronomical phenomena at all wavelengths. Note then that only half a century ago this was not yet possible, while now we have a thriving enterprise in space for many purposes; we will rave about that in Secs. 6.4 and 14.5, and Fig. 15.2 will show an example of such a spacecraft.

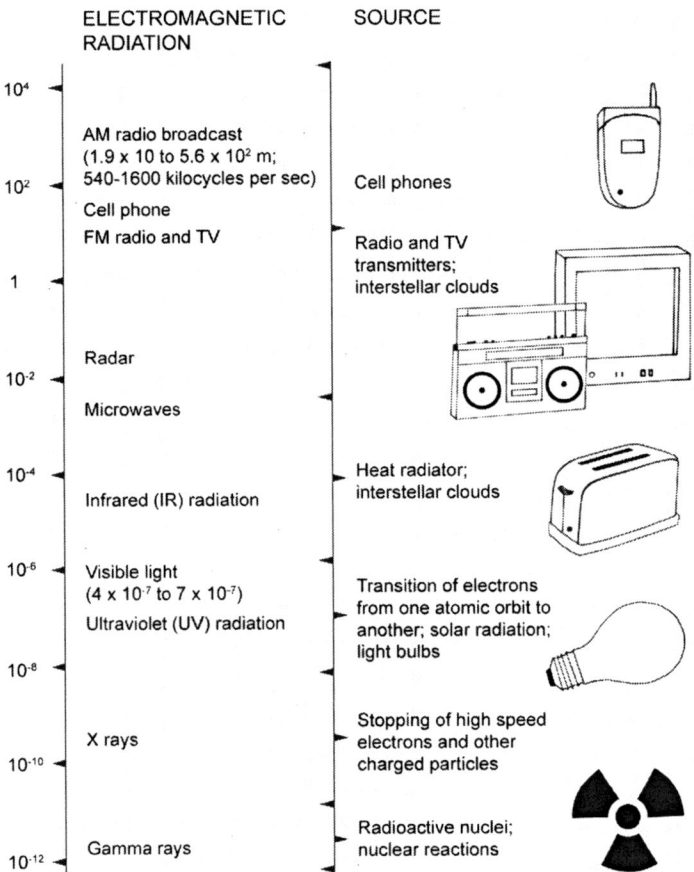

Fig. 1.6. The entire range of electromagnetic radiation. Note the small range of visible light. The wavelength numbers on the left are in meters.

The next figure shows the altitude levels at which half of the radiation from the sun, stars, etc. comes through in spite of ozone and oxygen absorptions in the far ultraviolet, and of carbon dioxide and water absorptions in the far infrared. Figure 1.6 also has sketches how we can observe with a variety of equipment the various radiations that the universe puts on display. It is not always necessary to use expensive spacecraft for going above 100 km, because a fraction of the radiations may come though to levels that rockets and even high-altitude balloons can reach. Instruments on rockets provide for some of the observations, as is shown, but they come down quickly again, so this is only for short observing times on a bright

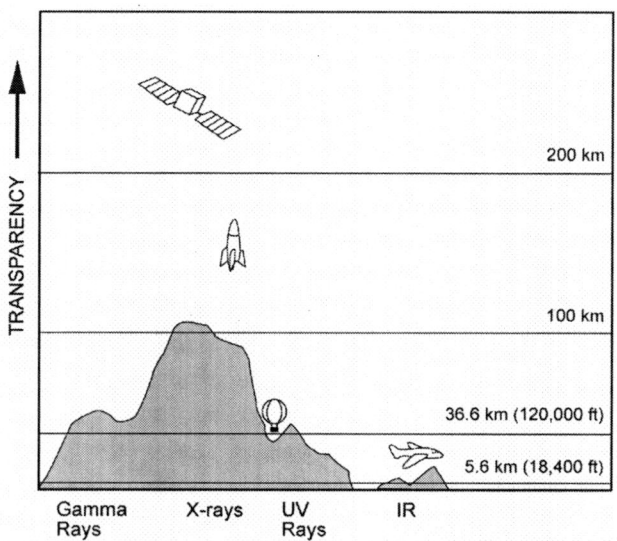

Fig. 1.7. From gamma radiation on the left, to long wavelengths on the right. The ordinate on the right gives altitudes. What looks like a mountain is a sketch of the absorption of incoming sunlight; the top altitudes of the profiles are for half of the radiation coming through the atmosphere.

object. The balloon is cheaper yet, and a balloon sketch indicates the narrow gap in between oxygen and ozone absorptions at 225 nm where unique observations were made by a University of Arizona crew with their own telescope and gondola floating at night near an altitude of 36.6 km (120,00 feet)[14]. At still shorter wavelengths the radiation is referred to as x-rays and, yet shorter, gamma rays. Our doctors and dentists use x-rays because these penetrate energetically through flesh and differently through bones and abscesses so that the effects can be used for analysis.

The sun warms the earth's land and oceans, and they re-radiate some of that heat back into space, but at a much lower frequency (longer wavelength) because the earth is much cooler than the sun. While the energetic sun radiates at a temperature of about 6000 K the earth does it near

300 K, a factor of 20 less. The wavelength of the sun's radiation happens to be near 6000 Angstroms (600 nm), and the wavelength of the earth's reflected heat occurs at that factor of 20 longer, 12,000 nm or 12 microns, deep in the low-energy range of the infrared. On the right in Fig. 1.7, we see the absorption of the earth's infrared radiation by CO_2 and H_2O; this occurs at about a third of the altitude of ozone so one can fly above it with telescopes in high-altitude jet aircraft. As much as half of all the light of the universe is in the infrared, so this is an important wavelength range for space research.

Back down in the atmosphere and on the surface of the earth we have the phenomenon of **global warming**. (We should not confuse it with the above ozone absorption, which is a quite different phenomenon, of different molecules, and much higher in the atmosphere.) The containment of heat has the name of **greenhouse effect** because that is what a greenhouse does, although it does it differently in part by not letting the warm air escape out, and that incidentally is why the inside of a car parked in the sunshine gets so hot. Our atmospheric greenhouse effect works by the absorption of the earth's infrared radiation, which then energizes the CO_2 and H_2O molecules to a higher temperature. This takes place by means of an intriguing mechanism using the electrical structure of the molecule, which functions like the small antenna inside a cell phone or radio or TV, vibrating and resonating with the incoming radiation, and in this case receiving and absorbing the terrestrial heat. The earth's atmosphere has vertical motion and mixing of the gases, such that the vertical temperature profile shifts to warmer, everywhere at the surface of the earth, and we speak of a *global warming* effect.

Before the very early earth had CO_2 in its atmosphere, it was radiating heat freely out into space, and its temperature at that time can be computed to have been about 255 K. That is well below 273 K so that any water on Earth was frozen; that famous 273-K level is in the Figure of Note 5 for this Chapter[5]. And when, much later, the Earth had evolved to have H_2O as we shall discuss in Sec. 11.1, the greenhouse effect came into place and the average temperature of the earth rose to the level near the present 288 K, well above freezing (it's all right to remember 288 as "near 300 K").

I therefore call this early process, rather anthropocentrically (Gk. anthropo = human being), the **good** greenhouse effect, because it made water and life on Earth possible. This is in contrast with the bad greenhouse effect, which is due to excessive CO_2 from industry and automobiles, while its resonance ability to absorb radiation then raises the temperature too much (Sec. 13.4). There are great fluctuations in the water-vapor content of the atmosphere, fun to observe in a desert climate for instance on a clear night, which tends to get chilly because the clouds are not there to trap the infrared radiation.

There also are intriguing effects on **sound waves** that we can listen for ourselves, such as the singing of the birds of course[15]. However, on that cold clear night, you may hear the sound of trains that travel unusually far in the distance. The clear night has the earth radiating its heat away into space, the air above the surface gets cold, dense and stable because vertical mixing stops, and the layering stratifies with the heaviest air being the lowest. The sound waves tend to move downward as they travel through a stratified layering, such that they are differentially retarded, and curved towards the heavier and more retarding lower layers. The opposite curvature effect can occur, but this is extremely rare. I once saw our dog barking at a distance of only about 30 meters, but I could not hear her because we were on the hot black lava terrain of the Cerro del Pinacate, south of the Arizona border. The hot black crust was irradiating the air above into being hot below and cooler above, which is rare because it is unstable. The sound waves were now curving **up**wards from their source, well above my head, into the higher cooler air that retarded the sound waves upwards, but my head did not get it.

Another special treat of observation that we celebrate near the times of sunset and sunrise is seeing the **shadow of the earth** (the sky must be clear[7]). Few people know about it, but once you have found it, you will have a great time pointing it out to others. It is simply the shadow of the earth projected against mountains and the air above them. The shadow may be seen increasing above the eastern horizon every evening, but only for about 10 minutes after sunset (and decreasing above the western horizon just before sunrise). To understand why it is visible from our perspective, imagine how the earth will cast a shadow when the sun is just below the horizon (the shadow is cast above the opposite horizon). It is darker above that horizon in a broad band of deep blue gloomy shade. How high up? Spread your four fingers, not including the thumb, on your arm stretched out towards the shadow - that gives the maximum height at which one can see the shadow. Above that height is a pink transition to the sunlit atmosphere, due to scattering by dust particles that are larger than the wavelength of light and therefore absorb especially the shorter wavelengths, leaving the longer pinkish ones to come to us. Now try a classical method for seeing such faint extended features easier, namely by bending far over sideways until the line through your eyes is at a **90-degree** angle with respect to normal. This breaks the monotony, a tiredness of the eyes, or a protective dulling of the retina senses. We have much more exploring to do, away from the earth.

2. THE EARTH'S ENVIRONS

2.1 Meteors and Meteorites

We leave our immediate environment and explore what surrounds the earth. We studied in Sec. 1.2 how the sun spews subatomic protons and electrons out in all directions (Fig. 5.1, below). This is the solar wind of Sec. 1.2, consisting of protons and electrons, in this case called "high-energy charged particles". This time, however, we must not use the word 'particle' for objects that are thousand million times larger than protons, the dust particles in interplanetary space. I shall avoid that confusion by using their other name, **interplanetary grains.** Comets and dust from colliding asteroids are the principal source for these grains; they are collected with high-altitude balloons, airplanes, and spacecraft. Even inter**stellar** grains seem to be among them, having come from much greater distances, and one studies their detailed differences and characteristics in laboratories[1].

There is enough of this dust floating in space, illuminated by sunlight such that our naked eyes can see them as a phenomenon, giving a faint glow in the sky near the horizon close to the sun after sunset and before sunrise. It is so faint, however, that it is distinguished best by using **averted vision**, also called **peripheral vision**. That is done by not staring at those regions directly, but rather by turning the head a little off, to the side, so that the "peripheral" sides of the retina will catch it, which are especially sensitive to very faint light. Imagine how peripheral vision must have favored our early forbears to notice some danger sneaking up on them from the sides, and this capability is still tested when applying for driving licenses, for the same reason. Anyway, this evening or morning glow looks like a triangle standing up on the horizon, best seen about an hour-and-a-half after sunset in the western sky, and before sunrise in the eastern sky. This is the **zodiacal light** because it is visible during the year successively along a belt of "zodiac"

Constellation figures that have mythological names ("zodiac" derives from the Greek word for carved figures). The planets also move along that belt, progressing through the same array of constellations in the **ecliptic plane** of the solar system. It is a different plane from that of the Milky Way. They are both seen best when they are well separated, around February and March for the evening western sky, September and October for the morning eastern sky. The faint hazy zodiacal light is fainter than the glow of the ensemble of stars in the Milky Way.

The earth sweeps quite a few interplanetary grains up as it travels in its orbit around the sun; these grains are **meteoroids before** their capture. Once they enter the atmosphere, they streak down the earth's atmosphere, and begin to glow due to collision with atmospheric molecules. Now they are **meteors,** "shooting stars", and a bright one is called a **fireball.** When the earth passes through a comet's tail, a large number of grains may enter our atmosphere so that we speak of a **meteor shower.**

An object larger than a fist does not burn up completely when it streaks through the atmosphere; when found on the earth it is called a **meteorite**[2]. Meteorite is the general name for any natural object that survived the fiery descent through the earth's atmosphere, but for an even larger body I specify that it is a comet or an asteroid. Anyway, the meteorite shows a burnt effect over the outer layer of the surface - it has been polished and blackened by our atmosphere - and that is how it can be identified even among the stones of the desert, which are much rougher.

In 1969, a huge meteorite fell near the Mexican village of Pueblito de Allende, it may have weighed as much as 4000 kilograms, and entered the atmosphere with a bright flash and frightening thunderclap, breaking up into thousands of pieces that were scattered over a wide area. It is usual practice to name a meteorite after a landmark or village nearest to the crash site, so this is "**Allende**", and it is famous because of its special composition, as we will see below and discuss further in Sec. 10.1.

Antarctica is also a famous location for meteoriticists because it preserves and yet discloses its meteorites by means of a curious mechanism. Following travel through the atmosphere and top surface, the fast object will sink into the layers of snow and ice buried deeply. However, the snow and ice layers sometimes drift gradually up a mountain slope and then the icy top layers will evaporate and the meteorite may become exposed to sunlight again. There they lie, near the base of the mountain, ready to be harvested, and our students go and have a great time collecting them, funded by federal grants of course because travel to and in Antarctica is expensive.

Meteorite searching and research began around the year 1800 when it was realized that they were coming from space – that was and still is a stirring realization and such hunting for meteorites and understanding them is sometimes called **the poor man's space program**. Here are samples of remote materials for the taking, having originated from long-ago times and far-away places, for us to study Nature's history and variety! Within meteorites such as Allende, material appears preserved from the early formation of the solar system; there are meteoritic samples that seem to have been irradiated by an exploding star during that formation, which was then probably caused by the explosion pushing matter together. Meteoriticists eagerly use these opportunities to learn about evolution by collecting and studying each of them as a potential time capsule[2]. We will now proceed to examine some of the larger bodies of our solar system.

2.2 Planets and Pluto

We will first consider the earth as an object among others, Earth as a Planet, and then ask the debated question for Pluto, "is it not a planet?" Table 2.1 is basic for making an overview of the solar system.

Table 2.1. Overview of the Solar System

Member	Distance (AU)	Size (x Earth)	Number of Satellites	Visible Characteristic
Sun	0	109	-	Hot Gas
Comets	$0-10^5$	small	?	Coma and Tails
Mercury	0.3-0.5	1/3	0	Rocky Surface
Venus	0.7	1	0	Cloud Deck
Earth	1.0	1	1	Cloudy Surface
Moon	1.0	1/4	0	Rocky Surface
Mars	1.5	1/2	2	Surface, a bit hazy
Asteroid Belt	2.2-3.3	small	yes	Rocky Surface
Jupiter	5.2	11	4+45	Cloud Deck
Saturn	9.5	9	1+50	Cloud Deck
Centaurs	5-50	small	yes	Snows and Rocks
Uranus	19	4	27	Cloud Deck
Neptune	30	4	1+12	Cloud Deck
Pluto	30-49	1/5	1+2	Methane Ice
Trans-Neptunians	32-98	small	yes	Rocky Surface
Eris	38-98	1/5	1	Methane Ice
The Oort Cloud	$10^4 - 10^5$	various	?	Comets

I rounded the numbers off to make them easier to remember for life, as behooves us for knowing our solar system. However, the distances of each object from the sun would cause a series of incomprehensibly large numbers if they were expressed in kilometers. We therefore switch scales, and express planetary distances in terms of the **astronomical unit (AU)**, where one AU is the *average distance of the earth from the center of the sun* (about 150 million kilometers, 1.5 x 10^8 km). To get a feel for the enormity of that number, compare it with the earth's radius of 6,000 km, which is smaller by a factor of 25,000 (precisely, can you see that?).

The planets that orbit closest to the sun - Mercury, Venus, Earth and Mars - are called **the terrestrial planets** because they have properties in common, somewhat like the earth, that distinguish them clearly from the outer planets. They are mostly solid masses, whereas the outer ones are much larger and consist mostly of lighter substances such as gases, and liquids deeper in.

All the planets, comets, and asteroids have elliptical orbits that are displayed in Sec. 4.1; Kepler's Laws[3] tells about them. The **explanation** of the orbits lies in the **curvature of space** that occurs near massive objects because of atomic action. There will be more discussion on this fascinating topic and on the essence of space, waves, and particle properties in Chapter 8. To get used to the idea, Fig. 2.1 is an illustration of how space is distorted by the mass of the sun.

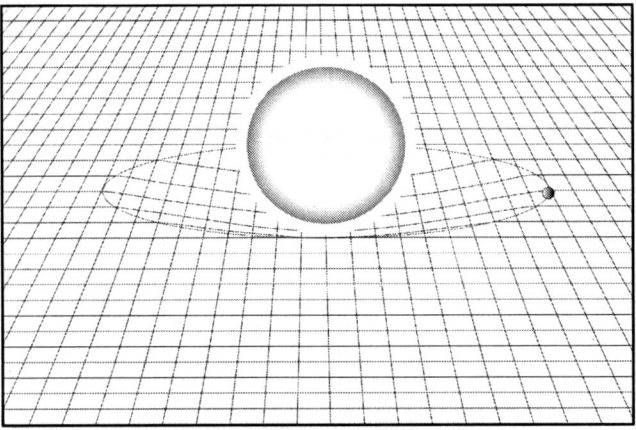

Fig. 2.1. The sun is at the center, and space nearby is sketched as it is distorted by the sun's mass. The dot indicates a planet as it follows the path shown.

The more massive the object and the closer it is, the more it distorts space. There appear to be three cases where the **space-curvature explanation** works especially well, and I will express this paragraph in those terms rather than in the usual referral to Kepler's Laws[3]. The second Pioneer mission to Jupiter was aimed into Jupiter's gravitational well (as seen for the sun in Fig.

2.1) in such a way that the spacecraft would be falling in, in 1974, be accelerated and have its direction in space changed in order to encounter Saturn on the other side of the solar system, five years later. It was the first time that such a maneuver was executed, and I heard its designers argue why Pioneer had arrived 20 seconds late (after a 5-year trip), but the technique became routine for missions to the outer solar system. Second, we will speak of a game of billiards in Sec. 4.6, how comets coming from the outer regions of the solar system may fall into the gravitational wells of Neptune, Uranus, Saturn, and finally into the deep one of Jupiter to travel all the way and become visible from Earth.

The third and most extreme case may be that of the demise of Comet Shoemaker - Levy 9. Orbits of celestial objects are usually determined by taking the gravitational effects of all planets into account. The planet Jupiter exerts the greatest influence by virtue of its large mass, after the sun of course, and the orbit of a comet or asteroid changes drastically when it ventures too close to Jupiter. The famous example of that was **Comet Shoemaker - Levy 9**, which on one of its orbits had come so close that Jupiter pulled it apart onto a straight line of twenty-some comets – and on the next orbit they returned and then collided with the upper atmosphere of Jupiter.

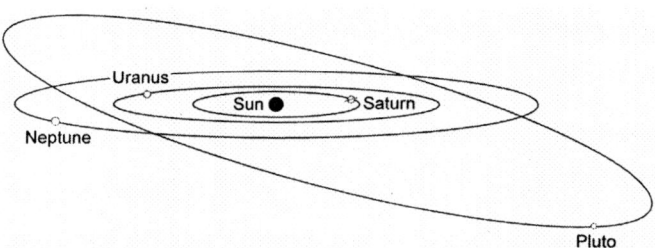

Fig. 2.2. The planets move in the plane of the ecliptic, but Pluto moves in a different plane, at an appreciable angle to the ecliptic, called "inclination".

Planets travel all in the same direction in their orbits about the sun, having that same direction of their rotation too, with the exception of Venus, Uranus, and Pluto. They also stay in nearly the same plane, the plane of the ecliptic, which we explained with the word zodiac in connection with the zodiacal light (Sec. 2.1). The word "ecliptic" itself comes from eclipses, which are likely to occur when two objects, and we the observers, are in that same plane and can thereby move in front of each other. It happens regularly in the case of **lunar eclipses** (the earth eclipsing sunlight on the moon), and much less frequently for **solar eclipses** (sunlight eclipsed by the moon). Mercury and Pluto are the exceptions in that their orbital planes lie at an appreciable angle to the ecliptic plane, especially Pluto (Fig. 2.2).

The reason for collecting all these regularities and exceptions is that this is essential information for understanding our solar system, and especially needed for investigation of its origin. It is also important to know how completely we know the populations of comets, asteroids and satellites, for instance for additional belts of cometary or asteroidal objects, near Uranus the **Centaurs,** and near Neptune the **Kuiper Belt** objects and the **Trans-Neptunian Objects** (TNOs). We pay special attention to the small bodies in the solar system because they participated in the formation of the sun, planets, and satellites of the solar system (and other stars probably have asteroids near them as well).

Searches are continuing for yet undiscovered comets, asteroids, and satellites anywhere, and for additional objects in the outer reaches of the solar system. There is also some hunting for any object with an orbit inside that of the Mercury orbit; it already has a name waiting for it, "Vulcan", but various searches have not found a sizable object there. In contrast, several programs and individual hunters find new asteroids on an hourly basis, comets are discovered at a rate of a few per month, and satellites less frequently, in special surveys for faint objects. When two numbers are listed in the fourth column of the Table, the first one is for the **exceptionally large satellites** that we will discuss in Sec. 4.4. Some 15% of the asteroids have at least one small satellite, sometimes two, while the activity of comets may split them into two or more components (Fig. 4.8, below).

What is a satellite? An object is a satellite if it orbits a planet rather than follows an orbit of its own around the sun. Earth's single one, **the moon,** is **atypical** in many respects such that the Earth-Moon system could be considered a double planet (and speaking of "moons" instead of "satellites" for other planets therefore seems inappropriate). The proportion of the moon's mass to that of the earth is much higher than that of the other satellites of the planets. The moon orbits Earth much more slowly than other satellites, and it is the only one doing that in the plane of the ecliptic rather than in the equatorial plane of its planet. In addition to all these special features, there is a haunting density problem too, which we will see in Sec. 10.4 together with a surprising origin theory for the Earth-Moon system to explain it away. We shall find the moon important for the rotational stability of the earth and that is essential for sustaining life, as we shall see when we wonder if we are alone in the universe (Sec. 15.2).

What is a planet? Planets are big objects orbiting the sun; the first six were bright enough to be known already in antiquity. For Uranus and Neptune, however, they could be found only after predictions of their locations had been derived from peculiarities in the locations of Jupiter and Saturn because of the gravity of the unknown planets, and then they were found. Pluto was too small to have any such gravitational attractions, so it was searched for and found in a special **sky survey** at the Lowell Observatory, in the early 1930s. Its small size and low mass were confirmed immediately after the discovery, but there was not much objection then to people calling it a planet, until recently.

Saturn's satellite Titan is larger than Mercury, and the four large satellites of Jupiter would qualify as planets if they had their own free orbits about the sun, but they orbit their planet, so they are satellites. **Pluto** is certainly not a gaseous or watery-icy giant as the four others are in that part of the solar system. The anomalies noted above in Pluto's orbit indicate an origin different from that of the four. Furthermore, Pluto now seems to be merely a representative of the newly known but large population of Kuiper-belt objects (Sec. 4.6).

Fig. 2.3. Pluto and its satellite Charon.

People everywhere are generally accustomed to calling Pluto a planet after all these years since the 1930s, and the **International Astronomical Union** (IAU) confirmed that in 2000; Pluto must continue to be a planet[4]. In 2006, however, the IAU assigned the problem of Pluto naming to two separate committees, which operated in secrecy. They surprised the Opening Ceremony of the IAU, which then had to vote on a new definition whereby all self-gravitating spherical objects would now be planets, including a satellite of Pluto seen in Fig. 2.3, and perhaps several asteroids, as well as Pluto itself.

I had published a pragmatic suggestion to stay with Pluto being a planet, and to extend that by naming new discoveries of objects larger than Pluto also as planets, with Eris, earlier called Xena, as an example[4]. My motive was the encouragement of astronomers to survey this frontier with the very best of equipment and telescopes – and they might do that if honored for the discovery of a planet. In any case, the IAU officials now had to come up with a new proposal in a hurry, before the Closing Ceremony. Their new proposal was to relegate Pluto, and those spherical asteroids and future discoveries as well, to **dwarf planet** status, and this passed with a majority vote; the principal criterion for a planet is to be massive enough to have largely swept up and thereby cleared its orbital zone. This definition captures the essential scientific principles[4]; the Soter reference has an illustration of the "new" solar system. That was it, and will be it until the IAU assembles again in 2009, and then anything may happen in face of a storm of protest from scientists and other people from all over the world. Pluto's status may be changed again. Ah, but all this notoriety is good for planetary science because of its wave of interest in small astronomical objects. Let us proceed with some big objects.

2.3 The Cold and Hot Surfaces of Moon and Mercury

The smallest detail on the lunar surface that can be resolved barely with telescopes on the earth is about four hundred meters; one then says that the telescope has a **resolution** of 400 m. Thanks to the Apollo astronauts' surface photography and their collection of samples, along with samples analyzed by robotic Soviet crawling vehicles, a resolution is obtained of better than a millimeter for selected parts of the lunar surface. We can say that we now know the lunar surface in fine detail, and there are plans by several countries for a return to lunar exploration, with landings by people from China, the U.S., and India.

The lunar surface shows craters due to impact of meteorites, asteroids, and comets. Observers are still debating if there are active or extinct **volcanoes** on the moon. It seems to me that there is no definitive proof of any active volcanoes, but a few colleagues disagree, and there certainly is evidence of outgassing, as we shall see.

The moon's surface is covered with a **regolith** (Gk. regos = blanket) a layer of dusty debris produced by the continual impacts of meteoroids and larger bodies (Fig. 2.4). The most frequent are, as on Earth, tiny meteoroids, but on the moon they encounter very little if any atmosphere and do not burn up. Because of their small impacts, the lunar surface shows fine detail at the highest resolution because the dust looks piled up in "fairy castle" structures. The great impacts of larger objects cause explosions that leave a crater and blast the surface material away, some of it over great distances. The surface soil is thereby turned over repeatedly, and such "impact gardening" is the cause of the regolith. Occasional asteroid and comet impacts must be occurring even now, but most of them came in two violently raging storms some 4.5 and 3.8 billion years ago; these are recognized from the rock samples that the astronauts brought back, and have since been dated by radiation techniques in laboratories.

Fig. 2.4. Apollo 12 astronaut Alan Bean standing next to Surveyor III, which stopped working long before. His own lander is in the background.

After the moon's formation about 4.5 billion years ago, a few large objects seem to have penetrated the surface, and molten material appears to have flooded their deep craters. These lava basins, flat and dark in contrast to the surrounding crust, looked like seas to the first eyes looking, so they were named **maria**, the Latin word for "seas" (the singular is **mare**)[5]. The other regions of crust are rougher and of different composition, which is brighter than the black lava, so that they appear different from the "seas", and therefore were called "highlands" a long time ago. The presence of water in those seas on the moon was believed tenaciously until the middle of the 19th Century, even though the well-known physicist/astronomer

Christiaan Huygens knew better 150 years earlier[6].

Fig. 2.5. Composite of images of the surface of Mercury.
The images were made by the Mariner-10 spacecraft.

The Apollo astronauts walked on the dusty, reddish-brown lunar regolith quite easily, their feet impressing the dust only a little more than sole deep. The lower gravitational force of the moon, one-sixth of that on Earth, allowed the astronauts to develop a **fast jumping gait**. Their footprints will remain for millions of years, little disturbed by the rare meteor and asteroid impacts and there is no wind to stir up the dust either since the moon has practically no atmosphere.

We discuss Mercury next because its surface looks similar to that of the moon[7]. Mercury and the moon are quite different inside, with a heavy metallic interior for Mercury, missing for the moon, indicating their different origins. Observationally, Mercury is difficult to study from here on Earth since the planet is so close to the sun, only visible to us occasionally, and then the glare from the sun makes it impossible to distinguish any surface details. The proximity to the sun also causes the daytime surface temperature to go up to more than **700 K**. Our main source of information about Mercury is therefore spacecraft, beginning with Mariner 10, which flew by the planet three times in the 1970s and imaged part of the surface with television cameras having resolution down to 0.1 km, 100 meters (Fig. 2.5). While the surface is thoroughly cratered, there are also cliffs or **scarps** up to hundreds of kilometers long and a few km high, not seen on the moon, which seem to show a shrinking of the whole planet, perhaps due to cooling after its original formation; there may, however, be a molten core[7].

Two spacecraft are to provide ample new observations from their orbiting about Mercury, namely MESSENGER (MErcury Surface, Space ENvironment, GEochemistry and Ranging), which has arrived, and, to come in 2013, "BepiColombo", named after the Italian space scientist who worked out special orbital maneuvers to facilitate imaging of large areas of the surface[8]. We now move to another hot surface.

2.4 The Hot and Cold Surfaces of Venus and Mars

The surface of Venus is veiled by a dense atmosphere[9]. The disk of the planet as seen from the earth is small, but it is observable with telescopes and instruments on Earth and on spacecraft. Images made in visible light show only the top of the opaque clouds (their high reflectivity helps to make Venus the brightest object in the morning or evening sky, after the moon). The surface however, looks like a murky brownish-reddish desert with rather smooth rocks and slabs, seemingly polished by fast sandy winds, but that became known only from the Soviet Venera spacecraft, the first to land on the surface, back in the 1970s. Theirs was a spectacular accomplishment, spacecraft surviving and sending information for some 40 - 90 minutes, from a surface temperature that is **higher than 700 K**.

Mountains and valleys of Venus have been mapped by radar from various spacecraft, augmenting early radar reconnaissance made from Earth, but now with more advanced radar imaging that has a higher resolution (Fig. 2.6). The word **radar** is from the combination of **ra**dio detecting and ranging. The range (distance) from the spacecraft to a certain feature on Venus' surface is obtained in principle, by measuring the time it takes between sending and receiving back the reflected radio signals. We know the velocity of electromagnetic signals much more precisely than our rounded-off number of 300,000 km/sec, and that precise number serves in such analyses. The simple principle is that twice the distance can be determined by observing the time between sending and receiving a signal. Although the police use the same method to measure the speed of an oncoming vehicle, on spacecraft the method is far more sophisticated because the measuring spacecraft is also moving, as is the object that it is measuring.

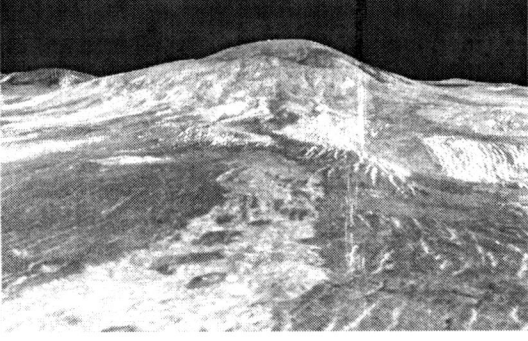

Fig. 2.6. One of Venus' volcanoes in a radar image made by the Magellan spacecraft.

However, detailed surface maps are the result. A large mountain complex on Venus is higher than Mount Everest, Maxwell Montes is its name in honor of physicist James Clerk Maxwell (1831 –1879); this is the only male name for features on Venus. There are large volcanoes, impact craters too, but only of large objects that managed to come through the thick atmosphere. Canyons are several kilometers deep, hundreds of kilometers wide, and a thousand kilometers long.

Regarding Mars, we know a lot about its surface from various orbiters, landers, and rovers[10] sent to explore the planet up-close, with more to come. The Martian landscape looks like a **brown-red desert** strewn with rocks and fine-grained sand. There is evidence of subterranean water ice, especially at the polar caps, which are white appearing regions that are imaged even from Earth. Landers and rovers have not identified any organic compounds, so far, but the hunt is on with several missions, also in orbit. The big question is whether **life forms** are unique to Earth, or if any primitive ones might be present on Mars (Fig. 2.7). Some interaction occurs with meteorites that seem to have come from Mars to Earth.

Fig. 2.7. The surface of Mars imaged by the camera on the Pathfinder spacecraft.

The variety of large Martian surface features offers a feast for the eyes as we can readily observe craters with many cuts and fractures, plains between the craters, and lava-flow patterns surrounding extinct volcanoes. Olympus Mons is **the largest volcano** in the solar system, dwarfing Mauna Loa on Hawaii, and Vallis Marineris is **the largest canyon** in the solar system, named after the Mariner spacecraft that made the early spectacular discoveries. That canyon is nearly 4000 km long, up to 250 km wide and at least 6 km deep, such that it dwarfs our own Grand Canyon which is "only" about 400 km long, 20 km wide, and 1.6 km deep. Small black holes are found, possibly the openings into caves.

Mars has two small **satellites**, Phobos and Deimos, named after fear as in phobia, and after deity in Deimos. Both have cratered surfaces that must have been caused by asteroids and comets doing their horrendous impacting. A dominant crater appears to have almost broken Phobos apart as can be seen from the long cracks ranging all over its surface. We shall next make a comparison with Earth for what we have learned thus far.

2.5 <u>The Surface of Earth by Comparison</u>

It seems likely that the earth was as **cratered** as the moon, but only relatively few impact features are recognizable today. It is, however, an active sport to find and identify the few impact craters and we presently know of nearly 200. Other impact sites may have been crushed away by the movement of tectonic plates, and/or may have been further eroded away by rain and weather.

In any case, the cratered lunar surface seems to show that asteroidal objects clobbered the terrestrial planets together in violent beginnings. That Mercury's surface looks similar to that of the moon we now understand, as an effect of similar impacts and formation, and of never having had **plate tectonics** and **an appreciable atmosphere**.

One can then understand that the present surface of the earth is not the original one. This makes the earth unique, evidenced by having seas and oceans and an involved and evolved flora and fauna, and especially oxygen and life. With the water and plate tectonics, both missing on Venus and Mars, nearly all the differences are clear as we study our inorganic and organic evolution. The movement of tectonic plates sculpts and chisels the earth, the plates colliding and drifting over hot spots cause mountains, subduction brings volcanoes and the lava wells up to form new crust - but the effects are also limited because they are not stagnant at one place, and the atmospheric erosion will smooth it all. For Moon, Mercury, Mars and Venus we find no significant movement on their surfaces, other than Mercury's contraction scarps, while the earth is an unstable battleground of tectonic activity with horizontal drift and lateral friction. The reason why Earth is the only one having plate tectonics seems connected with the abundance of liquid oceans **lubricating** the mobility of the plates, and so it is again the water that makes the difference. The Himalaya and the Tibetan Plateau show a detailed example of such lubrication[11]. The atmosphere of Earth is important for understanding our origins as well as for breathing the air, so let us study atmospheres and their mechanisms more in detail in the next chapter.

3. ATMOSPHERES OF THE TERRESTRIAL PLANETS

3.1 Motion and Outgassing of Molecules

The atmospheres of our moon and the planet Mercury are so tenuous that they are virtually non-existent, but there is some **outgassing** from the surface. It means literally that gas is coming out, this is clearly seen when it comes abundantly from volcanoes, but rocks and oceans expel gases too, albeit slowly, with molecules coming free from their stony[1] and watery prisons. Here is an experiment you can do if you are the first one in a swimming pool in the morning - move your hands slowly quietly just below the surface. Gas bubbles come jumping out!

The discoveries of outgassing for the moon and later for Mercury were celebrated events – a most surprising event - not only were gaseous molecules detected escaping from the surface, but also some **water frost** was found near the South Pole of the moon. That discovery was totally unexpected, and doubted at first, because the sunlit surface is hot. However, the Clementine spacecraft did make that discovery within deep craters near the South Pole where their bottoms are always in the shade, shielded from sunlight; the other necessary condition for the interior frosts to occur is that the lunar surface is mostly porous, not conducting heat from the sunlit rims and sides. In Arizona, near Sunset Crater with its porous lava fields, visitors used to be allowed to crawl into a small cave just below the hot surface, where ice could be seen year-round.

In order of increasing atmospheric thickness and density, we can line up Moon[1], Mercury[1], Mars, Pluto, Earth, and Venus; those of Moon, Mercury and Pluto depend strongly on local and temporary conditions. Next, in increasing order, are Neptune, Uranus, Saturn, and Jupiter[2]. Regarding Pluto, we know of the presence of a thin atmosphere of CO_2, but it may be there perhaps only at the time when the planet is closest to the sun, in its elliptical orbit, where the surface may be heated just barely enough to cause some outgassing. That is where Pluto is now. That thin atmosphere may **evaporate**, not to be observable again for nearly the orbital period of 250 years. That is why the "New Horizons" spacecraft is on its way to Pluto now, and it will arrive in 2015 when it is still sufficiently close to the sun to allow the study of its outgassing.

Molecules in an atmosphere are **in continuous motion in random directions as they keep colliding** with their neighbors. They do this more aggressively when they are aggravated by heat, with increasing atmospheric and radiated surface temperatures, and of course also when there are more molecules to contend with, when the density is higher. With increasing altitude there are fewer molecules coming from all directions to collide with each other. At extreme altitude, the molecules may be so far apart that some of their motions may not be deflected by another collision so that they simply keep going, away from their planet and out into space. Near that outer limit of an atmosphere, there is therefore a continuous **escape** of gases; in the case of the earth, we saw that this occurs at an altitude near 100 km. Near that altitude, the ultraviolet radiation from the sun breaks apart some of the electronic molecular bonds such that incomplete molecules, called **ions** and electrons are actually the ones that are escaping; please remember therefore that ions have a positive electrical charge, and that the electrons are negative.

If an atmosphere is maintaining itself, it must have a **balance** between demand and supply; a supply of gases from below will match the loss at the top of such venerable atmospheres. The escape of molecules from the earth's atmosphere is indeed in delicate balance with the outgassing from its rocks, oceans, and volcanoes. Mars does not have the latter two so its supply for the poor balance of its tenuous atmosphere comes only from the rocks, soil, and perhaps from molecules underneath the surface.

3.2 The Thin Atmosphere of Mars

Until the Space Age arrived in the 1940s, as is described in Sec. 14.5, we had to obtain our knowledge of the planets, satellites, comets, and asteroids from astronomers who observed them with their telescopes. There were only a few of them actually engaged and they had to overcome basic difficulties to study the fine detail because the observed planetary disks are small. Please get a feel for **telescopic viewing** of the planets at a Planetarium or Observatory near where you live.

Observing with Earth-based telescopes is troublesome due to the turbulent motion in our atmosphere, which smears the image by the resulting shimmering of the light coming through the atmosphere, and then it is difficult to see the fine detail of a planet in the vibrating image. We are familiar with the shimmering in the distance on a hot summer road, or over a fire, and even in a still pool one may see waves of parcels of warmed water rising and thereby disturbing the uniformity of the pool. A notorious case of this misinterpretation of fine detail occurred for Mars when astronomer **Percival Lowell** (1855-1916), older brother of poet Amy Lowell, reported seeing lines on the surface of Mars. Because the lines appeared to be straight, Lowell concluded that he had found evidence of canals, built by an advanced society in a life-supporting climate. We now understand better how the human brain has a tendency to connect features and line them up to see a straight line, as for a canal. Lowell's conclusion sounds outrageous now, but so little was known about Mars at the time. He may have been influenced by a preconception of Italian observers who had used the word "canale" earlier. However, soon there were astronomers with larger telescopes in steadier atmospheric conditions who could not verify the observations, and the idea fizzled.

The atmospheric pressure at the surface of Mars is only 1/150th that of Earth (a pressure of 1/150 "atmospheres", or 0.007 atm). It consists mostly of **carbon dioxide** (CO_2), with a few percent nitrogen and argon. The temperatures vary between near freezing sometimes near the equator and about -55 degrees centigrade (218 K) at the poles; the same temperature as outside of a commercial jet at high altitude.

Escape of atmospheric molecules is important to consider again for understanding the tenuous one that is left on Mars. Pictures of the surface show how flowing water is likely to have been present and seems to have sculpted **valleys and channels**. The giant volcanoes mentioned in Sec. 2.4 must have poured out an enormous amount of water as well as gas, lava, ash and dust, temporarily causing an atmosphere much thicker than the present. But when the volcanoes stopped erupting on this cold and cooling planet that lacks plate tectonics, the supply of molecules to the atmosphere halted, while the evaporation at the top of the atmosphere continued. The water was probably lost from Mars' surface in this manner, but some of it may still be present below the surface. We see something similar in Arizona where the small creeks and rivers can be dangerously roaring after a heavy rainfall during the monsoon, but most of the time the riverbeds look very dry. Underneath the surface, however, the river keeps flowing slowly and one sees bushes and trees in the riverbed thriving from the upwelling evaporation even during a drought in this the dry climate.

In summary, we understand that Mars was like Earth in its early evolution, but its atmosphere and water evaporated into space[2,3]. Mars may have undergone extreme **climate changes**, alike Earth with its Ice Ages, but on larger scales. Mars' warming and early thickening of the atmosphere presumably was the result of volcanic eruptions that temporarily changed the climate into one resembling Earth's. This may have occurred a few times in Martian history to explain the many observed geological features[3,4].

The Martian atmosphere above ground is now quite dry, but it also must still contain some **water vapor** because television cameras on landers have shown a frost on the spacecraft itself. Water-ice clouds on Mars that are formed during the night usually evaporate and disappear after sunrise, and small clouds form in the wake of high mountains. On Earth, the latter effect may be seen when the wind forces the air over a high mountain, and as it moves to higher altitude, there is expansion and cooling, which causes cloud formation in its wake. On the downwind side of mountaintops and ridges, the cloud formation may appear in the wind as a pattern of standing waves. Over the Martian polar caps of CO_2 frost, there may also be H_2O clouds, especially during the winter months.

However, such manifestations of water in the atmosphere are small by Earth's standard; Mars appears as a dry and cold **desert planet**, even if frozen water may be present below the surface. The winds are usually slow, but occasionally accelerating to cause **dust storms**, sometimes so severe or extensive that large parts of the surface vanish under dust clouds for months. There are distinctive patterns of dust flows left on the surface like those of the earth's sand dunes. Let us now look at the opposite of a tenuous atmosphere, namely the thick one of Venus.

3.3 The Dense Atmosphere of Venus

In *visible* light, no distinctive cloud features are seen for Venus because of a **droplet haze** in its upper atmosphere. But in the 1960's we were astonished to find that the droplets consisted of.... **sulfuric acid**, H_2SO_4 which is the same highly corrosive acid as is used in car batteries. Sulfuric acid is so vicious that if it is dropped on a leather shoe it will burn right through (Fig. 13.1 shows its extreme acidity). The discovery on Venus was made as the result of observations made from the earth and high-altitude balloons, measuring the polarization of sunlight reflected by Venus' clouds[5]. The observations were compared with computed models of such polarization, and a precise fit was obtained for spherical droplets a few microns in size, provided the droplets had the light-scattering characteristics of H_2SO_4.

Ultraviolet light makes it possible to observe dark markings in Venus' upper atmosphere, even from the earth, so that it becomes possible to study wind directions and velocities at high altitudes. The markings may be due to a haze of small particles of sulfur or sulfur dioxide produced by the breakup of the sulfuric acid by sunlight. That dark haze circulates around the planet once every 4 days, fast compared to the 243 days of the planet's rotation, and both are puzzling rates. The **slow rotation** of the planetary body is in an **opposite direction** to that of Earth and the other planets - this may be due to a big object hitting Venus early in its formation, but this is not a theory established in detailed modeling as was done for the origin of Earth and Moon (Sec. 10.4). The Venus Express is in orbit, and it studies particularly the upper atmosphere and its circulations.

Below the cloud layers, the atmosphere is not opaque but **murky reddish**, to the extent that landers on the surface of Venus have observed it to be gloomier than on an overcast day on Earth. The pressure at the surface of Venus is about 90 times what we are accustomed to on Earth. The most remarkable fact about the lower Venusian atmosphere is that it is incredibly hot, as we mentioned before and shall discuss further. Soviet spacecraft made the first descent through the atmosphere. It survived only some 40 - 90 minutes because of the heat, but that was a spectacular accomplishment as it was long enough to confirm the startling news. American spacecraft followed the Soviet ones to Venus and after only a few missions (compared to Mars missions), the planet had become well known. After a 15-year hiatus, the European Venus Express had arrived for more observations, to study the atmosphere at various wavelengths, particularly in the infrared to obtain penetration in order to determine cloud motions[6].

Where in our solar system is the worst weather nightmare? The surface of Venus must be the easy winner because its searing winds of a heavy atmosphere would cause you to fry instantly in the 900°F heat, or the gloomy reddish darkness would make you surely have nightmares that you were in hell without a drop of water to quench your thirst.

3.4 The Atmosphere of Earth by Comparison

The **distance from the sun** is the basic cause for the great differences between Mars, Earth, and Venus. Mars could retain only a small part of the water that had probably come from volcanoes and comets, but that would freeze out because there was not enough solar heating. Even ice evaporates, albeit slowly, and eventually almost all of the surface ice had disappeared; some of it may still be beneath the surface, and frosts are occasionally seen above. Because it is closer to the sun, Earth is warmer and larger than Mars, so that significantly more of the water vapor was kept long enough to bring about the good greenhouse effect we discussed already in Sec. 1.4 and will see time and again.

Greenhouse effects probably occurred at least a little on all three planets to begin with, such that the temperatures rose somewhat in their atmospheres. On the earth, it apparently increased enough for the water to remain liquid, making an atmosphere and plate tectonics possible; the latter may simply happen by lubricating the sliding of plates.

On Venus, closer yet to the sun, the solar heating was simply too strong. While the earth's greenhouse remained partially transparent, the Venusian atmosphere nearly closed and the temperature rose to its present high level of more than 700 K. The water that boiled off from the surface became too hot for even its vapor to stay. A small amount of water in the cooler upper atmosphere became bound to the sulfuric acid, resulting in its crystalline structure, which is denoted as $\mathbf{H_2SO_4 \cdot H_2O}$.

On Mars, the CO_2 remained frozen, trapped in the rocks below the surface as hydrocarbons, the atmosphere never became massive enough to sustain itself, and so it declined. Old and, now dry, riverbeds and rough channel terrain show that Mars did have water originally, but it must have evaporated, while some of it may be frozen and **trapped below the surface**. Even on Earth, much CO_2 is locked up in the rocks as hydrocarbons.

The meteorological conditions on Earth vary much more than on Venus and Mars. Venus never has a clear day because its cloud layers are too thick, while on Mars the atmosphere is too thin for more than occasional winds and dust storms to occur. But on Earth the diverse climates and the effects of weather are brought about by the **variation of its lands and oceans**, with winds driven by the differential heating and cooling of these land and water masses. The presence of water is unique for Earth, and its life is therefore unique as well.

The present atmosphere of the earth is composed mainly of nitrogen plus oxygen. That atmosphere is in striking contrast with those of our neighbors, who have mainly their CO_2. Something drastically different must have happened on Earth, and, indeed, we know that the difference is life. **Interplay of oxygen and life** occurred and a new type of atmosphere emerged, with oxygen being a necessary characteristic for life forms (nitrogen is less active). Studying the disparate conditions of the terrestrial planets shows boundaries for the conditions of life[7]. The earth formed at the right distance from the sun. If the earth had been in Venus' place, the temperatures would have evolved much too high for life. If it had been as far from the sun as Mars, it would have been too cold, and again life could not have evolved. This is sometimes called the Goldilocks Principle after a children's story about finding close fits. These are some of the considerations needed for our understanding whether or not we are alone in the universe, which we shall discuss in Chapter 15. First, we must complete our tour through space.

4. BEYOND MARS

4.1 The Asteroids

Figure 4.1 shows the spacing of orbits about the sun for Mercury, Venus, Earth, Mars, and Jupiter, to show the large gap between Mars and Jupiter, where another planet was therefore expected. The shaded area represents thousands of orbits in an **asteroid belt.** Note that there is a large

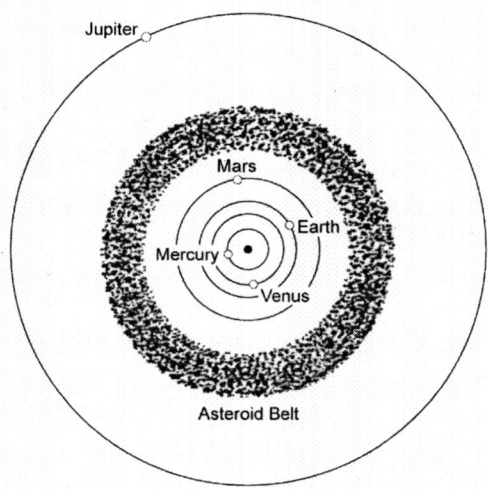

Fig. 4.1. A cross section through the asteroid belt.
The asteroids generally move not in the ecliptic plane
but at inclinations to it, as was shown for Pluto in Fig. 2.2.

gap between the orbits of Mars and Jupiter; the four inner orbits of the terrestrial planets seem tightly spaced, but a next one in that sequence seems missing. Would you not have expected a planet there? A group of astronomers in the late 18th Century did, so they held a conference in 1800 and decided to search for that planet[1]. Independent of that group, a Sicilian astronomer happened to be making a catalogue of stars, and he found a stellar image that was moving among the other stars! He knew that stars could not do that for they are so very far away; you can see the effect by an airplane in the distance that seems to move slowly even though we know it to have a high speed. He also knew about comets moving when they come close to Earth, but they have fuzzy images, not sharp like a star[1].

However, soon after that discovery, other such small bodies were found instead of the one large planet predicted between the orbits of Mars and Jupiter, and they were identified as **minor planets**. Nowadays they are more generally known as the **asteroids**[1,2], a word related to their star-like appearance, even when seen through a large telescope, instead of a disc for a planet ('aster' is Greek for "star"). That point-source aspect is still essential in distinguishing an asteroid from comets, which look fuzzy because of their outgassing into what is called a coma; we will discuss their behavior in Sec. 4.7. The first minor planet was named Ceres after a Roman goddess, and in 2006 it became a **dwarf planet** because of its spherical shape, as is explained for Pluto in Sec. 2.2.

Astronomers have not considered asteroids highly - they called them the **vermin of the skies** because of their motion making a trail and thereby sometimes spoiling a beautiful exposure. By the 1970s, however, the understanding had emerged that the small bodies in the solar system play a fundamental role in the formation of stellar and planetary systems, and in the 1980s that they had affected the evolution of life on Earth by a massive collision that eliminated the dinosaurs, giving the smaller bodies a better chance of survival.

Comet and asteroid observing has the special aspect that the objects are fugitive, seen to be moving with respect to the stars because they are so much closer to us; again, that is seen for airplanes too, appearing to move fast when flying by closely. The asteroids need a hunt or **"survey"** to be discovered, and then they need pursuit or **"follow up"** with many points observed along the orbits so that these can become well established. A special breed of astronomers was born which seems attracted to observing and studying these little fugitives - I call them asteroiders[3].

Asteroid discoveries had been accelerated from 1 in 1801, to 2000 officially recognized in 1975, and 4000 in 1985. After that, a surge in **asteroid discovery** came with electronic detection programs that search for dangerous objects such that in 1999 there were 10,000 and 48,000 in 2002,

300,000 in 2005..., with orbits determined for all of them after a considerable effort of following each object and determining precise positions at each epoch.

Table 4.1. Characteristics of Asteroids

Number in Catalogue	Name	Perihelion (AU)	Aphelion (AU)	Eccentricity	Inclination (degrees)	Diameter (km)	Type
1586	Icarus	0.2	2.0	0.83	23	2	U
2062	Aten	0.8	1.1	0.18	19	1	S
1	Ceres	2.6	3.0	0.08	11	952	C
106	Hecuba	3.0	3.4	0.07	4	70	S
624	Hektor	5.0	5.3	0.03	18	150x300	U
2060	Chiron	8.5	18.9	0.38	7	130	comet

Table 4.1 shows orbital and other characteristics for a few notable asteroids. The same orbital parameters apply to planets, asteroids, and comets and they therefore have general importance. In the columns one finds:

- A sequential **number** in the asteroid catalogue;
- The **name**, usually proposed by the discoverer;
- The asteroid's closest distance from the sun in its elliptical orbit, called **perihelion** (Greek "peri" = "near", "helios" = "sun");
- The greatest distance from the sun, the **aphelion** (Latin "apex" = "the highest point");
- **Eccentricity**, an increasing number for the deviation from a circle;
- The **inclination** angle between planes of ecliptic and asteroid's orbit (Fig. 2.2);
- **Diameter** of the asteroid in kilometers;
- For some indication of the **composition**, see below.

The six asteroids in Table 4.1 are in order of increasing perihelion distance. Icarus was the bold one in mythology, who wanted to fly and explore, approaching the sun ever closer, to the point of melting the wax holding the feathers to his wings, so he crashed but surely became the example for scientists to explore to the ultimate. The planetary Journal ICARUS is therefore so named, while Icarus' large eccentricity shows that it travels from close to the sun to well beyond the Mars orbit. Aten on the other hand stays close to Earth between 0.8 and 1.1 AU. Ceres and Hecuba are the only ones shown in the **main asteroid belt**, which lies primarily between 2.2 and 3.3 AU from the sun - numbers that are easy to remember for life. Ceres is the largest of the asteroids; it was discovered first and therefore has number 1 in the catalogue. Hektor is an example of "Trojan asteroids", objects as far away from the sun as Jupiter, and having a peculiar gravitational relationship with Jupiter. Chiron was the first discovered Centaur, named after the fabled half man and half horse creatures; observers discovered its outgassing activity as a comet much later. Like the words "planet" and "satellite," asteroid is something like a **catch-all term** for a variety of objects, they may have one or more satellites[4], and Fig. 4.2 shows one.

The naming of asteroids is a fun topic. The discoverer has the right of naming and that is so for so many that it got out of hand, after cats, dogs, and mistresses, which differs from the early days when the naming was after distinguished namesakes. For example, number 87 in the asteroid catalogue is Sylvia after the mother of Romulus, and her two satellites are Remus and Romulus. During the Cold War, we named an asteroid after Soviet physicist/dissident Andrei Sakharov, which may have improved his plight in house arrest and hunger strike because it drew great attention in the Soviet Union. Two KGB men even thanked me for doing it, and we had a vodka party to celebrate Sakharov.

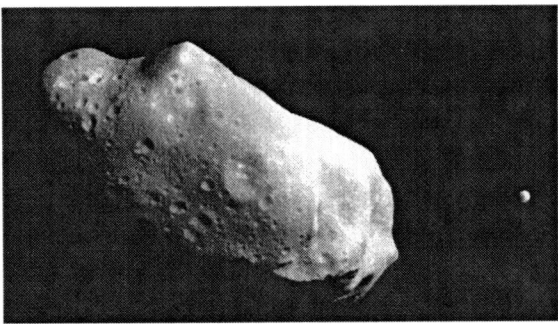

Fig. 4.2. Asteroid Ida and its satellite, Dactyl, imaged from the spacecraft Galileo while it was flying by.

The shown composition classifications are for asteroids only. Spectroscopic observations of sunlight reflected by their solid surfaces suggest that one should classify them into distinctly different types based on their composition. However, since the sunlight does not penetrate the top layer, the classification is for a **surface** composition, not for the material in the interior, which may differ. C-type objects consist of materials such as Carbon; they are quite dark in appearance, most likely due to hydrocarbons in their composition, and some show evidence of the water in crystalline form. Asteroids of type S are a mixture of stony and silicate materials; some have been found to be their loosely configured **rubble piles**. The rarer types include M objects, which probably are the metallic cores of large asteroids that originally differentiated (Sec. 1.3), but later lost their outer layers through collisions[4]. Finally, there are D, which stands for **d**ark red (sometimes thought to be due to organic matter), and U for **u**nknown composition.

The population in the asteroid belt seems to be "stirred", perhaps by a few large ones with exceptional orbits; the stirring refers to their traveling by each other at high speed, namely 18,000 km/hr on average. What if they have a **collision** at such speeds?! Asteroids often do collide with one another and then they both get shattered into fragments, but only if they are of comparable size. If one is much smaller it merely makes an impact crater. Because of so much collisional grinding, there are millions of asteroids having 1 km diameter in the asteroid belt alone, their numbers depending on size, with Ceres the largest, and the numbers increase as the asteroids become smaller. The effect is simply demonstrated by making a collision of a **hammer with a brick**, for there will be a few large fragments and evermore of increasingly smaller sizes.

Finally, **Jupiter's gravity** may throw some of the collision fragments helter-skelter towards other regions of the solar system. Eventually some of these asteroids may be caught by the gravitational pull of the earth or one of the other planets, and end their own lives in a collision, and so may we in the process (Sec. 12.4).

This does not mean that the asteroid belt is a crowded place – there are many asteroids, but they are thinly distributed because there is so much **space** in the belt. Half a dozen spacecraft have already traversed the belt from Earth to Jupiter without ever coming even within sight of a sizable asteroid; special targeting and maneuvering are needed for getting a space mission to come close to an asteroid. This was learned the hard way with the first mission to Jupiter originally announced as the Asteroid and Jupiter Mission, until more detailed study and modeling showed that it was not so.

Here comes a curious and important observation, namely that the S-type asteroids occur mostly in the inner parts of the asteroid belt between the orbits of Mars and Jupiter, while the C types dwell towards its outer regions. This shows the **transition** from the stone and silicate materials that the terrestrial planets mostly consist of, to the lighter materials such as hydrogen and helium that are abundant in the outer parts of the solar system. The DAWN spacecraft, using ion propulsion, is to orbit Vesta, representatitve of the S-types, and then Ceres, representative of the C types[4]. This is an intricate introduction to the next Section.

4.2 Jupiter and Saturn

Our understanding of the origin of the solar system is largely thanks to the fact that Jupiter and Saturn retained the lighter gases, the original hydrogen and helium from the early stages of the origin of the universe, while the radiation of the sun had removed them from the domain of the terrestrial planets. Such early evolution explains the qualitative difference of the terrestrial planets with Jupiter and Saturn. They are by far the largest planets in the solar system, 11 and 9 times the size of Earth with more than 90% of the mass of all the planets combined[5,6]. They both have **deep atmospheres**, which again are of compressible gases so that they show a great increase in density towards the center of the globe. The weight of the immense atmospheres causes the pressure to become exceedingly high towards the interior. The temperature then rises because of the greater activity so that the molecules collide harder and more frequently and that brings higher pressure as well as temperature. The effects become so extreme in the outer planets that the gas begins to have properties of a hot and heavy liquid, like a stiff gel. Do not be fooled by the word "liquid", for it is not fluid like water on Earth.

Deeper inward yet, it gets even more interesting as the molecules are pressed even more closely together and colliding so hard that some electrons are knocked out of their bonds with the atomic nuclei. The atom or group of atoms that are made are called **ions** (Gk. ienai, to go, implying action, I guess, because an ion is aggressive), each with a positive electrical charge because one or more electrons, the negative charges, are missing. The material is **ionized**; we met the condition before when we introduced some such conditions inside smaller Earth. Anyway, the positive charges of the ions and the negative charges of the electrons cause aggressive electrical properties. Remember from the discussion in Sec. 1.2 that when electric charges move, they cause magnetic properties as well as the electric ones, electromagnetic effects. The moving charges again cause magnetic fields and radiation belts, **Van Allen Belts,** for Neptune, Uranus, Saturn, and especially Jupiter; their effects are much stronger and farther reaching than in the case of Earth.

Jupiter and Saturn are so massive that their bodies are still settling gravitationally, long after their formation and compilation 4.5×10^9 years ago. The heavier substances are still **differentiating**, we met that concept for Earth and metallic asteroids, but now it is the heavier helium sinking through the hydrogen. This settling causes the central compression to increase and thereby increases also the internal collisions so that heat and its radiation appear, more energy emerges from the planet than it receives from the sun - Jupiter and Saturn radiate about two or three times more heat than they receive. Still though, their temperatures do not go up to millions of degrees so nuclear energy generation does not occur as it does in stars. They are still planets since they are not radiating out by far as energetically as the stars do.

From Earth we see Jupiter and Saturn generally yellowish; to show contrasts, astronomers and space engineers exaggerate with red colors, but only Jupiter's red spot is somewhat reddish[5]. One cannot speak of a **surface** for a gaseous atmosphere, of course, but when in popular writing one uses the word, it refers to the top of clouds that actually may be floating at various depths. Imaging with telescopes on Earth, and especially closer up on spacecraft flying by or in orbit of the planets, reveals a fabulous world of whirls and eddies showing spectacular cloud and storm systems on the outer planets. The velocities of the clouds can be measured from the motions of these features, and particularly near the equatorial regions, high-velocity **jet streams** are thus observed.

Fig. 4.3. Jupiter with its Great Red Spot and other cloud formations. The black dot is the shadow of satellite Io.

The **Great Red Spot** is the outstanding feature of Jupiter (Fig. 4.3). In the north-south direction, it has about the diameter of the earth, and twice that in the east-west direction. Because the storm is so enormous, it persists over a long time, hundreds of years, compared to the lifetimes of the relatively small hurricanes on Earth that are only a week or so. In the atmosphere of Saturn[6], the spacecraft Cassini has shown a less obvious feature but nearly of the same size deeper down, along with other giant storms also at these depths.

The **rings** are the outstanding features of Saturn (Fig. 4.4). A small telescope will already show them, while some of the fainter rings and satellites may be observed from the earth with special imaging techniques on suitable telescopes. The great progress in understanding the rings and other features in the outer solar system came, however, through observations with spacecraft flying by or, better yet, when they can be maneuvered to orbit the planet and then stay for a long time like Galilei did for Jupiter and Cassini is still doing for Saturn. We now know of extensive systems of rings and small satellites for all four outer planets.

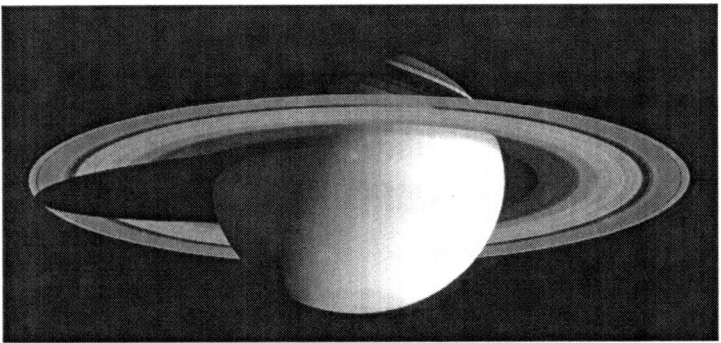

Fig. 4.4. Saturn with its bright rings.

Spacecraft that fly by or are in orbit of the distant planets provide the most detailed knowledge because they get close to them. They can also observe over a range of phase angles as we see for the moon, of **scattering angles** for the sunlight to observing instrument, from which the characteristics of the substance are determined - different materials scatter light differently into various directions. Another unbeatable advantage of observing in space is that **all wavelengths** are observable, as we illustrated in Fig. 1.7.

Because of the variation in distance and in latitude (north-south direction), gravity profiles inside the planets can be determined, and those profiles brought the discovery of **heavy cores**, which are deep inside and therefore otherwise invisible.

The flybys in the outer solar system began with the **Pioneer** spacecraft by Jupiter in 1974 and Saturn in 1979, followed by the larger and better-equipped **Voyager** for its Grand Tour by Jupiter, Saturn, Uranus, and Neptune, and by several of their ring and satellite systems. Two more large spacecraft followed later, namely **Galileo** to Jupiter, especially to its satellites, and **Cassini** to Saturn carrying the **Huygens Probe,** which was delivered at Titan (Sec. 4.4). On we go with our own Grand Tour, off to Uranus and Neptune.

4.3 Uranus and Neptune are Different

The differences of the four outer planets are again dependent on their distance from the sun; Uranus and Neptune may have formed later than Jupiter and Saturn, as we shall discuss in Sec. 10.4. All four have radiation belts, but with interesting individual characteristics. The appearances of the outer two are also quite different, without the spectacular belts and zones, of Jupiter particularly. Uranus and Neptune show a mostly uniform blue haze, and their overall body **densities** are higher, which observationally supports the mentioning of oceans, below[7,8].

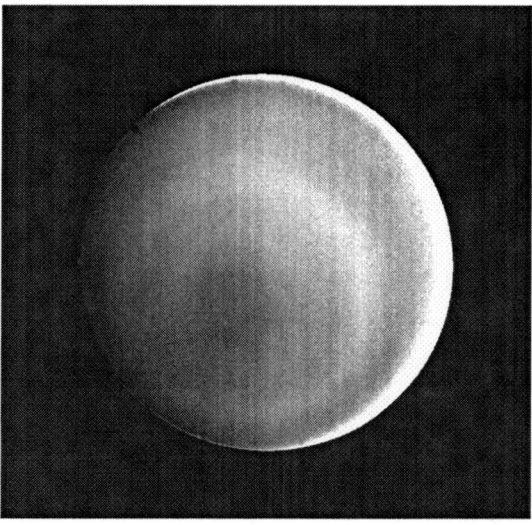

Fig. 4.5. Uranus showing, at this time, its polar region.

The **light blue** appearance of Uranus, and it seems even **deeper blue** for Neptune, are presumably due to methane in the cold upper atmosphere.

Its CH_4 molecular structure and size are such that it resonates with the red part of the solar spectrum, absorbing that, while the bluer light is free to reflect out. Water and ammonia may be also be present, but the temperatures in the upper atmosphere are low such that the vapors near the top will be frozen out from the gas phase and their crystals will grow, sink, become pressurized, heated and melted into the liquid phase, and that is how both planets are theoretically modeled to have **oceans**. However, these two outer planets will not be known much better until special space missions are sent to study them, for which no great plans have come to the fore as yet.

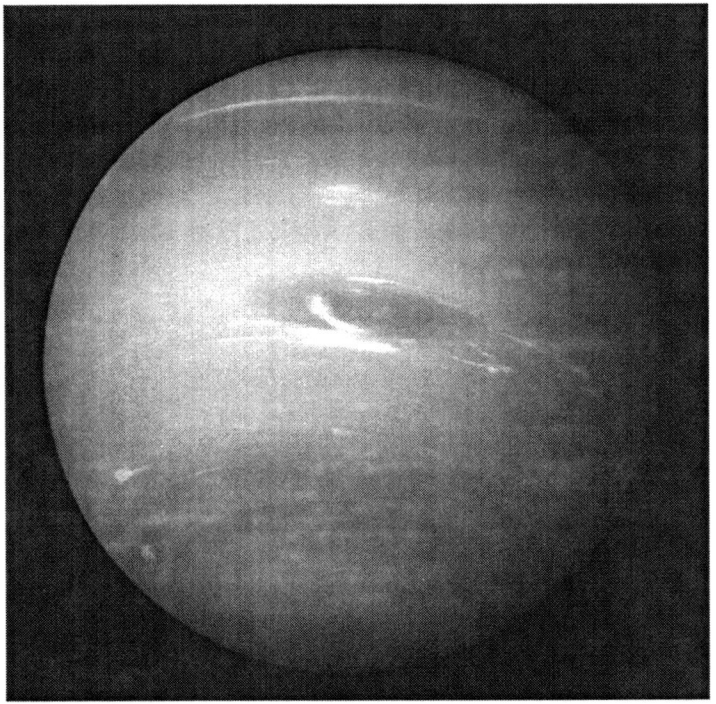

Fig. 4.6. Neptune with atmospheric features.

The outstanding feature of Uranus is an unusual orientation of its rotation axis[7]. It is nowhere near perpendicular to the plane of the ecliptic like the sun's, nor poised at an intermediate angle like the earth's rotation axis at 23.5 degrees, but **tipped to about ninety degrees** and thereby lying nearly in the plane of the ecliptic. We presently observe a polar aspect (Fig. 4.5), and we must therefore be careful in comparing that with Neptune, which shows the more normal nearly equatorial aspect (Fig. 4.6). The cause of Uranus' orientation remains a mystery. The answer may lie in terms of a collision with a large object close to the time of the planet's formation, when huge objects may still have been available for such an otherwise rare encounter. There has been no modeling with a detailed theory yet, as was done for the moon and Earth (Sec. 10.4).

The outstanding feature of Neptune[8] is the planet's **deep blue color**, caused by the methane cloud layer in its upper atmosphere being deeper than for Uranus. We also see a few large spots, which allow the study of strong wind velocities. Other than that, we know little about Neptune. This is not the case for the next six objects.

4.4 Titan, Triton and the Galilean Satellites

Six of the satellites[9] orbiting the outer planets are about as large as the moon or even Mercury; they would have been called planets if they had been in space by themselves. They are **Io, Europa, Ganymede, and Callisto** of Jupiter, Titan of Saturn, and Triton of Neptune, which can be observed with telescopes on Earth, but they are not much larger than pinpoints because of their large distances. Spaceflight made the difference so that we now know them well.

The early history of the four Jovian satellites is fascinating, with astronomer and physicist **Galileo Galilei** being the first to see them moving near Jupiter. That happened in 1610, with one of the first telescopes, which did not yet yield resolution as well as that of our binoculars, but he did see them and noticed that they were **orbiting Jupiter!** That was an astonishing finding, at odds with the emphasis of the Church on Man and Earth being at the center of the Universe, and one should therefore expect everything to revolve around Earth. With the announcement of his discovery, Galileo ran directly into the Inquisition, a frightening Council of Cardinals, who supervised and controlled the adherence to church doctrine. Galileo's invitation for the cardinals to come look through his telescope was not accepted. In the end, he agreed to keep quiet about his findings, knowing that the recalcitrant philosopher Giordano Bruno had been burned alive in 1600 for speaking out irreverently against Church doctrine; on a square in Rome, Campo de Fiori, one can still see a marker tile of that gruesome place.

The four largest of Jupiter are now called the **Galilean satellites** in his honor, and so was the **Galileo** spacecraft that went there to study them. It made many detailed discoveries, particularly of electron beams connecting Jupiter and Io, and a magnetic field around Ganymede, which protects it from high-energy Jovian particles, alike the earth's Van Allen belts fend off the solar wind. After dozens of orbits around Jupiter, its mission completed and its control fuel depleted, Galileo was sent crashing into Jupiter to avoid a possible collision with Europa making pollution there that could affect a future mission.

Back to the much earlier Voyager days, a junior assistant had spotted what looked like a **volcanic plume** in one of the spacecraft images of the surface of **Io**. This was a surprise! The explanation was that the four satellites and Jupiter have complex tidal effects, alike we saw a little already for Moon and Sun tugging at the oceans and surface of Earth. The tugging causes a continuous friction of matter inside Io, a constant "kneading", which heats the interior to the point of partial melting. The heated molten material expands and bursts forth in volcanic activity. We learned of half a dozen giant **volcanoes** spewing sulfur and SO_2 onto the surface of Io, some appearing and others decaying, on and off.

The same kneading process occurs inside the other Galilean satellites, and such heating explains that there is at least some water in liquid form. An ocean may lie beneath the icy/rocky surface of even the outermost Galilean satellite **Callisto**, while cracked fields of water ice are seen clearly on the surfaces of **Ganymede** and especially on **Europa**[9]. The Galileo mission had already found a water ocean underneath the ice floes of Europa, and planetary scientists are dreaming of and scheming for a new mission for exploration with submarines. This would be done robotically because human explorers are not likely to visit the Galilean satellites, as the Jovian radiation systems seem too strong and deadly.

The Cassini spacecraft was designed for extensive observations of Saturn and its satellites, and for dropping off the Huygens probe through the atmosphere of **Titan** down to its surface. The names are in honor of Jean Dominique Cassini (1625–1712) and Christiaan Huygens (1629–1697), both pioneers in the understanding of Saturn's rings. The Huygens probe landed with a parachute on Titan as scheduled, carrying instruments for making spectroscopic and other observations of the atmosphere and the surface. The main spacecraft Cassini is to complete some forty flybys of Titan with powerful instruments in operation. The atmosphere is a 300-km thick hydrocarbon, but mostly nitrogen-orange smog, nearly opaque to observations of the surface from the outside. The Huygens probe obtained images all the way down and on the surface showing what appears to be a liquid sea or dry bed of methane, with coasts and hills, and tributary shapes that seem to have let **liquids streaming into the methane sea.** Titan appears to be in many ways comparable to Earth, with impact, volcanic and atmospheric features, of the colder methane in this case instead of our water because of the great distance from solar heating.

Triton orbits Neptune in the direction opposite to the general direction of the orbits of most other satellites[8]. It is quite a contrary exception, but there are a few other satellites like this one and they are explained similarly, to be objects that formed independently of their planet, Neptune in the case of Triton. It was formed somewhere else and had a history of its own, having been bumped around by other free objects, thereby getting a rotation all of its own. Eventually, however, Neptune captured Triton with its gravity. But it would forever show its own original direction of orbit, contrary to that of most other satellites that had been formed from and dragged around by a dense nebula surrounding their mother planet.

The final encounter of the 12-year interplanetary journey of the last of the two Voyagers was with Triton[8]. It became a fitting and culminating ending to the Voyager missions, after so many discoveries and fascinations! The surface features of Triton were **a summary of several surprises** at other satellites encountered earlier in the missions. Active plumes visible in Triton's thin nitrogen atmosphere suggest geysers; with hot liquid forced up to the solid ice surface, forming features that froze quickly into their snowy beauty now admired on Triton. The reflecting power over the surface is high but variable due to nitrogen and methane snows, particularly in a polar cap that lies over what appears to be ice emerging along stretched ridges and mare-type plains. After revealing all these riches, Voyager continued on its flight away from the solar system, off into deep space. We still have to look at a variety of objects in the system.

4.5 Rings and Small Satellites

The outer solar system is a wondrous world of snow and ice, and of satellites[9] and rings[10] around the planets. Saturn's main rings may look like solid plates from Earth, but Voyager's intricate observations show that they actually consist of thousands of thin and narrow rings. It has also become clear how a satellite could not have formed that close to the planet because of differential gravitation: at each minutely different distance from Saturn, an object has its own elliptical orbit with a different velocity around the planet. Objects even at slightly different distance from Saturn will therefore be sliding by each other, not coagulating into a solid ring plate or even into a satellite – their differential velocities are too high, their collisions too violent. Saturn's rings thereby consist of a multitude of **thin rings** of icy grains, snowballs, or ice-covered rocks having sizes from meters to microns

(10^{-6} meter).

Details of Saturn's ring structures and their interactions with satellites and the magnetosphere are in the forefront of research. We also know for the three other large planets of many thin and/or tenuous rings, inviting further study and comparison. While Saturn has five rings, one of them broader, outside of the closely bright system we discussed, Uranus has eleven thin rings, Neptune two narrow rings and a broader one, and Jupiter has three faint rings[11]. The discovery of **so many rings** was one of the triumphs of the Voyager program.

As for the **small satellites**, we knew of 80 in 2001, 113 two years later, and there are 134 in Table 2.1 (in 2007), which gives us a feel for the rates of discovery[11]; this is done with large Earth-based telescopes that are also used to follow up for improving the orbits. The diameters of satellites range down to a few kilometers for the smallest detectable size. The satellites consist mostly of snowy substances, the same as must always have dominated the outer parts of the solar system; at the great distance from the sun, because the lighter, volatile elements could survive where the solar heat had little effect.

Impacts of comets and asteroids make the surfaces of the satellites look heavily **cratered**. There were surprises for individual satellites since some of them showed craters of large impacts that nearly broke them up[9], while small satellites may occur close to rings, interacting with them. In the case of Saturn's F ring, which was a Pioneer discovery, two little moons now seem to act as shepherds to the ring particles, keeping them in their present place by a complex gravitational interplay. We knew nothing of these intricate situations until the Voyager flights in the 1970s. What fun it must have been to possess such powerful tools and to derive such detailed understandings! For us there is more to come, in the environs of Uranus and Neptune.

4.6 <u>Centaurs and Trans-Neptunian Objects</u>

Neptune's orbit marked the outer edge of the known solar system for a long time, but there were searches for other planets, and the discovery of Pluto resulted in 1930, we have mentioned Ires in Sec. 2.2, and there still are searches in the hope of locating other distant planets. We know of new groupings of fairly large objects, namely the Centaurs near Uranus, and the trans-Neptunian objects (TNOs). It begins to appear that TNOs with nearly circular orbits do not occur beyond 50 AU, which indicates that 50 AU may have been the **original outer boundary** of the solar system and of the nebula from which the solar system formed. Discoveries **are** being made far beyond 50 AU, but these are of TNOs with exceptionally elongated elliptical orbits.

Centaurs are objects near the orbit of Uranus, but with a wide range, from 5 AU when closest to the sun till 50 AU when farthest; they seem to be unstable wanderers. The first one, Chiron[12], was discovered in 1977, it obtained the usual identification as an asteroid, but cometary activity was eventually observed. It took 15 years before another object was found at that distance, but now we hear regularly of new discoveries, and the name of this grouping is Centaurs after the mythological man/horse monsters; Chiron was a son of Saturn, a grandson of Uranus. Many objects may belong to this category, perhaps as many as in the main asteroid belt between the orbits of Mars and Jupiter.

The Centaurs seem to be intermediate objects that are transferring from the ones farther out yet to the regions farther in. **Chance encounters** with any planet may change an orbit drastically, perhaps throwing it out of the solar system or bringing it in towards the inner reaches. Eventually, some of them may become comets in the inner parts of the solar system.

Beyond Neptune is a populous region that we referred to already as trans-Neptunian objects. We begin to understand them better as their discoveries increase and analyses improve. The nomenclature becomes more detailed, with **Kuiper Belt** objects of low inclination, and **scattered disk** TNOs farther out having high inclinations[13]. The interest in these two populations and Centaurs is high because they occur in a new frontier, an unexplored part of the solar system, helping us understand the origin and evolution of these populations and thereby of the solar system as a whole.

Pluto is considered a Kuiper Belt object, we mentioned that in Sec. 2.2. It is much smaller than Neptune, but massive enough to show a thin atmosphere. Pluto has a satellite, called Charon, which seems exceptionally large for a satellite, so that an encounter of the two rather than an accretive origin has been proposed[14]. There also are two small satellites. A proposal to especially study and fly a "New Horizons" mission to Pluto was at first not funded by NASA for budgetary reasons, but the refusal brought enough protest such that the spacecraft is on the way, will fly by Pluto in July 2015, and may even proceed into the Oort Cloud. Let us look at that cloud and at comets in general.

4.7 The Oort Cloud and Activity of Comets

Continuing our journey through space outward, on the final stretch in the solar system, we have to travel an enormous distance from the Trans-Neptunian Objects within 50 AU, to 10,000 AU distance from the sun where we encounter the next known population, that of the Oort Cloud of comets[15, 16]. They can even occur within a 100,000 (10^5) AU diameter region about the sun, but not much farther because there they approach the **stability fringe** of the solar system. Farther away from the sun its gravity may not be strong enough to hold an object in a solar orbit against the attraction of other stars or massive clouds in space. Figure 4.7 gives a sketch of the geometry with the stars projected on the ecliptic plane, which in this case is also the plane of the paper (the depth of the third dimension is therefore not shown).

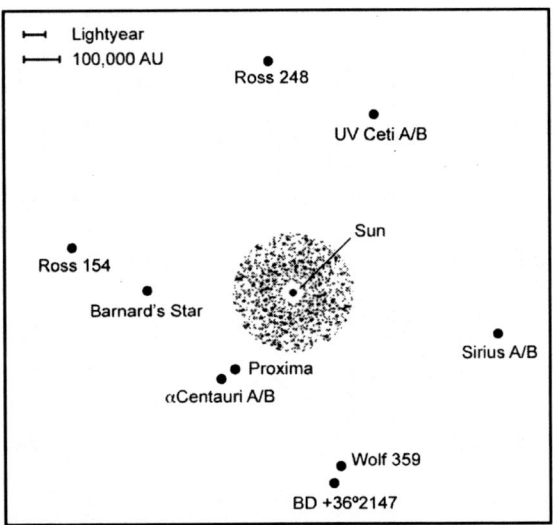

Fig. 4.7. Comets of the Oort Cloud projected on the ecliptic plane. Again, the comets have large inclinations, away from the ecliptic plane.

Note the peculiar effect that on this scale - on which another civilization would view us - most of the solar system is the Oort Cloud; it is not a small outer ring as one might wrongly believe. The figure also indicates that the nearest stars are about four lightyears away at this time. However, that distance may change at any time since all stars are always in motion, and so are the interstellar molecular clouds that are floating among them. The gravitational effects of these massive objects can therefore

attract, distort, and perturb the Oort Cloud, with the result that comets will be lost into interstellar space while others may be hurled inward into the solar system.

We have discussed already several times how in the outer regions far from the sun it is too cold for volatiles to evaporate away, yet cold enough to keep the frozen snows and ices. Wherever loose dusty and rocky materials can stick together to grow into comets, now or in the past, the snows, ices and gases will be mixed in, and vice versa; the comets are collected and assembled together as **dirty snowballs**[16]. They are loosely stuck together because the gravity of a small cometary body is weak. This is not to say that comets are small - they may have a nucleus of several hundred kilometers, and then the configuration becomes gravitationally stronger. An interesting phenomenon is that particles may freeze together, "cold-welded"; this effect, incidentally, is the fright of mechanical engineers who design moving components for spacecraft!

Computer modeling shows that such assembly of comets has happened in the early stages of the formation of the solar system, and that these dirty snowballs grew to sizes of up to hundreds of kilometers. These objects may be the original cores of comets and that formation process probably occurred originally in the regions of the outer planets Uranus and Neptune. We have already alluded to the possibility that the birth of Uranus and Neptune was a special event, late in the formation of the whole solar system. During the comet formation by coagulation and assembly of masses of dust, rocks, snows and ices, the ensemble gradually came under the growing gravitational influence of Uranus and Neptune, which may have thrown them out gravitationally to those 10^4 - 10^5 AU distances, where they remain for a long time in cold storage. Remember that any one of them thrown out father than about 10^5 AU would be lost to a life of wandering in interstellar space. On the other hand, any thrown not as far as about 10^4 AU would return to the inner parts of the solar system by gravitational effects of the planets. We saw already that gravity of a passing star or molecular cloud might occasionally perturb the outer comets that are near $10^4 - 10^5$ AU, causing them either to leave the solar system, or to travel back inwards. All these comets from the various distances might then be passing by Neptune and other planets, which would perturb them by their gravitation to come further in, or be thrown in other directions perhaps even away from the solar system. I call it a **gravitational billiards** game.

As for the theory, Jan Hendrik Oort (1900-1992) studied the orbits of comets and determined when they were observed for the first time, when they were called "new comets", and he noticed from their orbital parameters the great distances from which they had come, $10^4 - 10^5$ AU. Details of the origin of the solar system were still unknown in 1950, but Oort knew of the early ideas that it had been born from a solar nebula. He surmised that comets then formed from the dusty and snowy cloudiness near Uranus and Neptune, and that the gravitational perturbations by the planets had probably moved them to the outer reaches. He therefore added the above theory to earlier studies made by others already in the 19th Century[17]. There has been no observational verification of the Oort Cloud; it is all theory because the comets are too small so that at such a distance they are too faint and unobservable until we send a **spacecraft** among them. Study of comets from spacecraft began with an international flotilla flying by the nucleus of Comet Halley in March 1986.

Professor Oort made his contribution in the form of a **paper** in a professional Journal, but he did that while teaching a class, because he was a very busy scientist who had little free time to do it otherwise. I had asked his permission to be there as an undergraduate among half a dozen graduate students, and he proved to be a quiet speaker, absorbed in his thoughts, so that we stood around him in front of the blackboard, fascinated and sometimes contributing to his ideas. It was a delight to participate, and a unique opportunity to see science in the making[17].

The most distant comets in the meantime were orbiting the sun in stable trajectories at speeds as low as a few tens of kilometers per hour and not necessarily in the plane of the ecliptic but with all possible **inclinations**, which are angles between the planes of their orbits and the plane of the ecliptic (Fig. 2.2). The disturbances by stars and clouds would have occurred randomly, throwing them in all directions and Oort had derived that from the great variety in the inclinations for the new comets.

These comets are at such great distance from the radiation of the sun that their temperatures are close to absolute zero. Later, after perturbation by distant stars and molecular clouds, they gradually approach the sun and they will warm up. The ices melt first, next the liquids start evaporating, and finally the gases heat up and expand. When the pressure of these expanding gases becomes high enough, they find or force fissures in the dirty snowball to burst out as gaseous jets like those coming from a jet engine but much larger and more violent, as in small volcanoes. As the gases burst out from the inside, they carry the dust grains along with them and that is how we get to see the dusty hazy environs of the nucleus called a **coma**, which may reach out hundreds of thousands of kilometers, as a noisy whirling world of gas and dust roaring away from the nucleus. I wrote the word 'noisy' purposely because there may be enough gas to propagate the sound waves.

In completely empty space this is not so - we cannot hear the sun over our great vacuous distance, even though there must be quite a racket in gaseous outbursts and explosions near the solar body.

The **Deep Impact** mission crashed a heavy impactor into a comet, while the main spacecraft imaged the results; exquisite on-board maneuvering made it all possible[18]. A large dust cloud, brightly illuminated by sunlight, augmented the coma. It seems that the comet had a tenuous fluffy make-up, reminding us of talcum powder that is held together by the low gravity of the small nucleus without any sticking by cohesiveness of the material.

Fig. 4.8. Spitzer Space Telescope image of Comet Schwassman-Wachmann 3, breaking up. The fragment is too large, and ions are too small, to be affected by solar radiation pressure, so that it stays on the straight ion tail.

In a direction away from a comet near the earth, we generally see a tail of grains, the **dust tail**, as a curved feature due to the interaction of sunlight that keeps pushing by the light against the grains, radiation pressure. Another and different tail consists of molecules which by the solar radiation become ionized, *i.e.* stripped of some electrons, and that tail is straight because the molecules are smaller than the grains and they do not have enough cross-section to be affected by the radiation pressure of the sun. This one is the **ion tail**, seen in Fig. 4.8 even to be reaching forward of the comet. Spectroscopy reveals a wonderful assortment of ions such as H_2O^+, CO_2^+ and the spectroscopists have a great time observing and interpreting the composition of each individual comet.

When comets of low inclination from the Kuiper Belt finally reach Jupiter, they will be caught by the strong gravity and come into a new orbit, now determined by Jupiter as well as by the sun. In such a complex but **periodic orbit,** most of these comets cannot make many more than some **500 approaches** to the solar heat, because each time it loses gas and dust, as well as the ices and snows that might still be deep inside. Those losses go on until all volatiles are gone and then there is no more activity - the comet is dead.

Here are two examples of comets that are still alive, still seen with coma and tails. The name they generally receive is determined differently from the case for asteroids, for comets it is **after their discoverer** and so we have Comet Encke, which was quite active more than 200 years ago, but may have only a hundred years left before its activity will cease. Another famous one is Comet Halley, named after the astronomer who computed its orbit and had found that it has a period of 76 years (it is also aging but with such a long period, it is good for many more than 500 orbits). Sure enough, appearances occurred in 1986, 1910, 1834 and in years long before that, while the next one now will happen in 2062. Will you be around? The writer Mark Twain (1835 – 1910) was not to see it in 1910 and not in 1835 of course, and so one speaks of it sometimes as "Mark Twain's comet".

5. THE STARS

5.1 Effects of Pressure and Temperature

We have looked at people, the earth and other planets and at the most distant parts of the solar system. That exploration included the planet Jupiter with its excess radiation, but not enough to produce nuclear energy as stars do, but now we do proceed to the sun and other stars. They consist largely of **hydrogen and helium**, as does Jupiter, and a small star is actually rather close in size to Jupiter; you can see in Table 2.1 that even the sun, an average star, is only 109/11, close to 10 times larger. However, the volume and mass depend on the third power of the size, so it is $(106/11)^3$ larger, nearly a factor of a thousand.

The stars are spheres in an intricate balance of their attractive gravitation pulling the gas towards the center, with enormous pressures and radiation, pushing in all directions. The stars are indeed so much more massive than Jupiter that the pressure and temperature exceed a boundary where **nuclear** processes begin; the stars shine, radiating their massive powers.

In order for us to understand what happens at the highest temperatures inside massive stars, we need to consider the progression of what happens when **pressure** is increased. To begin with, we have already discussed in Sec. 3.1 how a gas can be so tenuous that some molecules are bounced off to escape from the top of our atmosphere. Next, we came down to the surface of the earth, where the weight and compression of all the atmospheric molecules causes our comfortable atmosphere, at an amount and a pressure of what we call one atmosphere (1 atm). We cannot see molecules for the simple reason that they are too small, but we can feel them by waving a hand vigorously, or sitting on a fast motorbike. The molecules are not quiescent, even in a quiet atmosphere, but bounce around among each other and they are constantly bumped towards the next collision. This **energetic activity brings heat**, causing the temperature of about 300 K where we live.

To continue the temperature progression of compressible gases, we looked at Jupiter with its much deeper atmosphere and we found the greater pressure bringing properties of a liquid. Actually, that was not a surprise because even for our laboratories we can buy liquid oxygen, nitrogen etc., which come in heavy steel cylinders because of the pressure. The cylinders must not fall, or the neck might break off and the cylinder would instantly be a violent rocket! Deep inside Jupiter, at greater-yet pressures, some **break-up of the molecules** into ions and electrons occurs, and their positive and negative charges make electromagnetic effects and Van Allen Belts possible.

But for stars, with their thousand-times greater volumes and masses, we must imagine much higher temperatures than in Jupiter as we proceed inward towards the stellar center and there we get to see the break-up of **the atoms**. At the supreme pressures and temperatures deep down into the sun, the atoms will be completely separated by their violent mutual collisions into nuclei and electrons. However, enormously higher pressures and temperatures now cause the break-up of **the nuclei** into subatomic components. Towards the centers of stars the stresses are violent enough and the collisions among the nuclei fierce enough so that even the strongly held together atomic nuclei break up. We are now in the domain of **nuclear energy generation** that makes the sun and the stars shine.

5.2 The Sun

We are fortunate to have the sun, a typical star - not the largest, not the smallest - near enough and therefore seen large enough in the sky so that we can examine small parts of the surface with special **solar** telescopes[1,2,3]. Stars do not have a solid surface for they are some form of gas albeit with a steep trend to greater pressures and temperatures inward towards their interiors. However, there is indeed a sudden change that looks like a surface, and that occurs not at great depth. This topic will be important to understand clearly when we discuss the age of the universe of 400,000 years (Sec. 7.5), so it is discussed here in some detail.

Deep inside the sun, intricate nuclear processes generate **photons**, waves of radiation - imagine them as small packages of energy made deep inside the sun and are now moving outward. That is not straightforward since the nuclei and electrons bump the photons into all directions, and not often in the outward direction. The going is difficult because the emerging light is knocked around repeatedly, sideways and even backwards – this process is called **multiple scattering**. It takes as long as a million years from the 15-million-degree center of the sun to reach a level where the temperature is 6000 K. There, the density of the sun's material is at a remarkable level.

At 6000 K, the density has decreased enough for the nuclei and electrons to have just the right spaciousness for combining back into complete atoms again. And so they combine. The main point reached here is that the nuclei and electrons are not free anymore to do their scattering of the light, knocking the photons around into all directions. At the 6000-K temperature level, the photons are suddenly **free to surge outward**, in between and through the newly completed and therefore no longer scattering atoms.

Now consider our viewing from the opposite direction, when we are observing the sun from the outside inward. At first we are clearly looking through that unimpeded outer atmosphere; the light has no problem proceeding because the nuclei and electrons are bound into atoms, not yet free to do any scattering of our light. Suddenly, at that threshold of 6000 K, our viewing runs into a wall of scattering by now-free electrons and nuclei. Our viewing suddenly encounters the layers of multiple scattering. That "wall" looks like a surface, and so we speak of the **surface** of the sun. Other stars may have different masses by which the discontinuity in the scattering occurs at different "surface temperatures" and these are the ones given in the abscissa of the diagram in Fig. 5.7, below.

Our solar telescopes do not observe the surface as being smooth; it looks a little like a boiling kettle but it is more like a violently exploding cauldron at temperatures sixty times higher than in our kettles. The heat is bubbling up from the interior by **convection**, but the temperature differences are so large over the surface that we see contrasts in a variety of **relatively dark features** and **sunspots** (Fig. 5.1). It is a much more violent convection than in Figs. 1.1 and 1.2. Detailed observation with special equipment on solar telescopes has taught the solar astronomers that **magnetic fields** are involved, like those of the earth but of much greater strengths. The high temperature has broken the material down into positive nuclei and negative electrons so that electromagnetism can occur and the sun's magnetic forces are having a field day. Loops and twists occur near sunspots, allowing the study of the solar fields in intriguing detail[4]. It is no wonder that the solar astronomers are devoted specialists who get to enjoy it all.

On the other hand, the protons and electrons of the solar wind we encountered in Sec. 1.2, come from special processes in the **outer** parts of the sun, in the **corona**. The hot outer corona is a faint extension of the sun; Fig. 5.1 shows a loop flaring out from the sun, and that is a part of the corona.

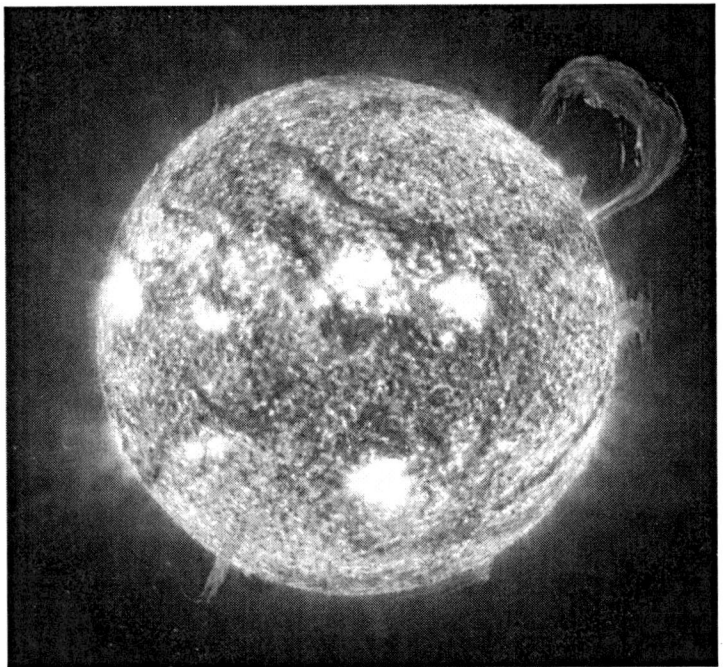

Fig. 5.1. Image of the sun made by the Solar and Heliospheric Observatory (SOHO) spacecraft.

The best observing of the corona is done with special equipment during **solar eclipses**, when the moon appears exactly in front of the sun. Since Sun and Moon happen to have the same **apparent** size as seen from Earth, and since they move at different apparent rates, the eclipse occurs for a few minutes only, and it happens only rarely because the Earth and Moon do not travel in exactly the same plane. Only once in a few years is there an opportunity to witness this event, if one is willing to travel to faraway places and to take a chance with the local weather; solar astronomers pursue eclipses for special studies and experiments on the corona, which seems to vary all the time. People should try to experience a solar eclipse because it is an experience of a lifetime; it gets dark during totality, animals react to the sudden nightfall, and the coronal appearances are unforgettable.

Astronomers use a variety of tools to gain information on the various radiations coming from stars, nebulae, comets, and asteroids; for the sun there are solar telescopes, also in space, such as **Stereo A**, *a*head of the earth and **Stereo B**, *b*ehind the earth in its orbit, and their data are combined to yield a three dimensional perspective. A general technique is by observing and analyzing the **spectrum**, which displays the brightness in different colors, at different wavelengths. The separation of the colors may be achieved in a **spectrometer** incorporating a prism or some other optical device. A detailed spectrum of a star shows a pattern of sharp black lines due to **absorption** of the light in its outer layers at specified wavelengths by various elements. The same antenna-resonance mechanism for making such sharp absorption lines we mentioned for cell phones, and for CO_2 in the greenhouse effect. In the case of a star, the properties of the atom are tuned in because their sizes are near the wavelength of the radiation so the atomic nucleus vibrates and resonates, with the radiation and absorbs it. The structural characteristics of each atom tune the interactions precisely, so that the wavelength and color have sharp precision, making clearly seen absorption lines in the spectrum; extra bright lines occur oppositely from **emission** (extra generation of energy) by hot gases (Fig. 5.2).

Fig. 5.2. Spectral emission lines:
A. Hydrogen; B. Sodium; C. Oxygen

Each element has a distinct pattern of these lines; hydrogen, for instance, has a well-known set of lines and each set has a Greek letter, so we speak of "H alpha" (H_α), "H beta" and so forth. **Spectroscopists** study the spectra and interpret them in terms of stellar compositions, temperatures, motions, types and constitutions.

5.3 The Variety of Stars

Stars may be smaller or larger than the sun, but because of their great distances, they all still look just like points - point-sources of light - even with the highest telescopic magnification, all, except the sun because of its proximity. The effect of distance on perception of size is easy to visualize by comparing the size of an airplane when it flies overhead to its appearance as nearly a speck when flying far in the distance. Eventually, the airplane vanishes because it has become too faint in reflected sunlight, and stars vanish later because they have their own illumination. Still though, their light will also become fainter with distance. Because of the distance, though, no size is seen; there is a single pinpoint of light. When the single beam encounters the turbulence of the earth's atmosphere, it may move in and out of the range of the small pupil openings of our eyes, and we say that **the star twinkles**. The visible planets, on the other hand, do not twinkle because they are disks, small but still having some extent of the disk. Therefore, when one part of that extended disk twinkles *out*, other parts twinkle *into* our eyes' pupils, so that on average there is a steady amount of light. Do check this by comparing with a star nearby on Venus, Mars, Jupiter, and Saturn when they are not too low in the sky. Mercury is too close to the glare of the sun, and Uranus and Neptune are too faint.

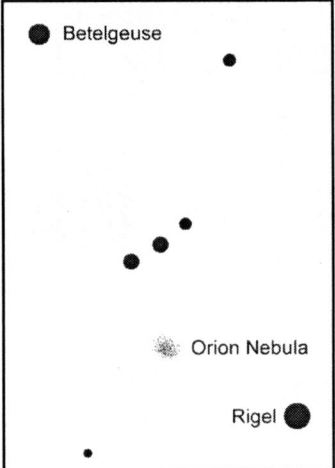

Fig. 5.3. Orion, as seen from the Northern Hemisphere.

There are some 10^{11} stars in the wide expanse of our galaxy, but only a few thousand of them are close enough and bright enough to be visible to the naked eye at anyone time during a clear night. Fainter stars become visible with the help of telescopes, long-exposure photography, and the faintest with modern electronic TV-type observing at the largest of telescopes. With the naked eye, we see areas of the night sky, and we notice apparent groupings of stars called **constellations**, which have time-honored mythological names given by our forebears. So we may speak of Ursa Major, Big Bear, a part of which is now known as the "Big Dipper." Another famous grouping is Orion with its three stars and two wider-spaced ones above and below, and its "Orion Nebula" indicated in Fig. 5.3, where stars are being born.

However, the stars we see together in constellations are usually **not together at all** - they only appear to be neighbors from our perspective on Earth. The problem is that we lack the perception of depth, so these stars are usually at quite different distances and therefore in reality not related or grouped together. For example, Rigel and Betelgeuse we see in Fig. 5.4 are at distances from us as different as 800 and 600 lightyears (see Table 5.1, below); oddly enough, the more distant Rigel appears the brightest of the two because of its much greater intrinsic brightness, "luminosity."

We are entering the enormous domain of galaxies and our universe so we will need to switch scales for sizes and a distance far greater than the Astronomical Unit. We have the distance that **light travels in a year**. One of the essential numbers to simply remember is for the speed of light of 300,000 km/sec - and you must realize and appreciate what a fabulous speed that is – seven-and-a-half times in a second around the earth (if we could make a beam of light do that). The **lightyear** name is confusing name - it is not a time, it is a distance. You can derive its 10^{13} km yourself, starting with 300,000 km/sec and with the number of seconds in a year. Rounding the numbers off gives a surprisingly close answer: 3×10^5 kilometers per second *times* 4×10^3 seconds per hour *times* 2×10^1 hours per day *times* 4×10^2 days per year. Multiply the numbers to 96, rounded of is 10^2, add all exponents to 10^{13} km, while the precise answer is 9.46×10^{12}.

Back to these stars drifting apart like all other bodies in the universe - because everything moves in the universe - so that the configuration of stars making up Orion may suggest something completely different to an observer looking up at the night sky say 100,000 years from now. Why so many years from now? The comparison with the airplane is useful again to understand motion and **speed** at a distance. The plane appears to move fast when flying overhead - we speak of high angular velocity - but slowly when far away. And so it is for the stars - the universe is **alive with motion**, but this is not perceived by us because of the great distances that are involved. Incidentally, here is a nice example of how we fool ourselves as observers: an incoming plane that is large will seem to fly slowly, but this is so only because we use its large size as a measuring stick for its speed - all jet planes land at about the same speed, near 200 km/hr.

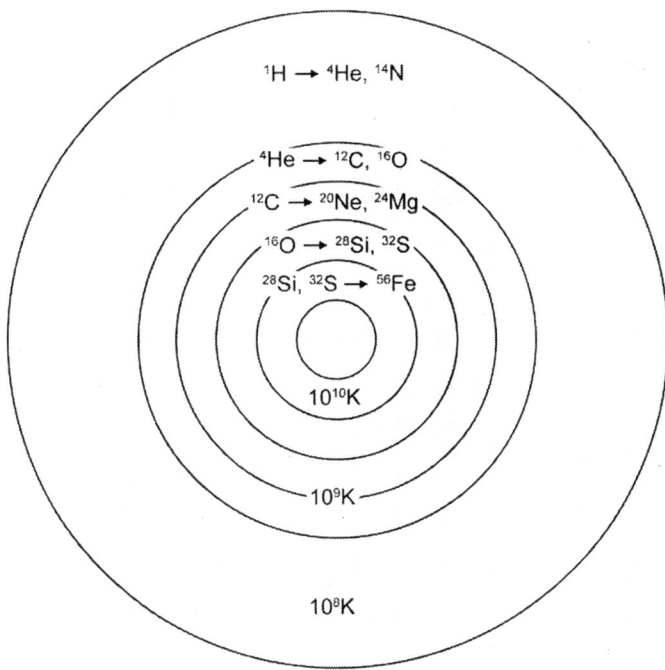

Fig. 5.4. A sketch of natural selection, showing how various increasingly heavier atomic nuclei are formed under increasingly more energetic conditions inside a star.

The essential difference between stars is that they differ greatly in their mass, the amount of material that is inside them. When they are more massive, there is greater activity because the bigger mass - larger over-all gravity - presses down harder so that the interior pressure and temperature are higher. Figure 5.4 is a classic we will use over-and-over, because it shows what happens in the **formation of heavier elements** in a star that has a mass of 20 suns, 20 solar masses. Note how the temperatures increase towards the center, because of the increasing pressure, and how therefore the nuclear reactions become more complex. The results of increased reactions and interactions are heavier nuclei such as of nitrogen, carbon, oxygen up to iron (Fe); the interiors of stars are giant thermonuclear furnaces that fuse the subatomic particles into greater combinations from those of hydrogen into heavier elements. We shall discuss the fundamental concepts later, while here we continue with our descriptions of the various types of stars.

Fig. 5.5. An open cluster. This one is known as the Pleiades, named after the seven daughters of the Titan called Atlas. Some of the dust from which the stars were produced is still visible in scattered light.

About half of the stars occur as pairs, **binary stars** orbiting around a common center of gravity, while others are single, or they may belong to multiple systems in which the stars move about each other in complex ways. Still larger groupings may occur in **open clusters,** which have the stars distributed rather widely as is seen in Fig. 5.5, while **globular clusters** are dense assemblies of large numbers of stars concentrated towards the center so they indeed look rounded as is seen in Fig. 5.6.

The open-cluster stars have spread apart since their common birth. Their direction and speed of motion are measured on photographs of say 70 years later, and by counting backward, their place of birth and, more importantly, the **age** of the cluster and its stars is determined. This a simple but clever technique so we know specifically that open clusters are young, and therefore their stars are relatively young compared to the age of the sun, which is nearly 5 billion years (5×10^9 y).

A basic tool in astronomy is called the Hertzsprung-Russell (HR) diagram, named after two astronomers who developed it separately, one in Europe, and the other in the U.S. The **HR diagram** is an indispensable tool for understanding various stellar types and their evolutions. Computations,

Fig. 5.6. A globular cluster is usually much older, without dust and young stars, but there are exceptions.

in astrophysics, of the structure and characteristics of stars are its mainstay - spectra, temperatures, and radiation output come from the physical effects at various depths in the star, with the results shown in Fig. 5.4. The vertical axis gives a measure of the star's total output of radiation, its intrinsic brightness, or **luminosity**. The labels of the horizontal axes are for color, type of spectrum, or with the temperature of the "surface", defined in the previous section. Even with the naked eye, we can see some color, red or blue stars; this is best done when it is not so dark and our vision is still with the cones of the retina for color vision, rather than with the rods for faint nighttime perception. Anyway, the left and right parts of the HR diagram

(Fig. 5.7) are for blue and red, connected with the temperature, and that is intuitively obvious, as we know that a flame of low temperature is reddish, while a very hot one is nearly white or perhaps even bluish.

Diagonally through the HR diagram, we see the **Main Sequence**, with our sun near its middle, at the center of the diagram. The sun's detailed

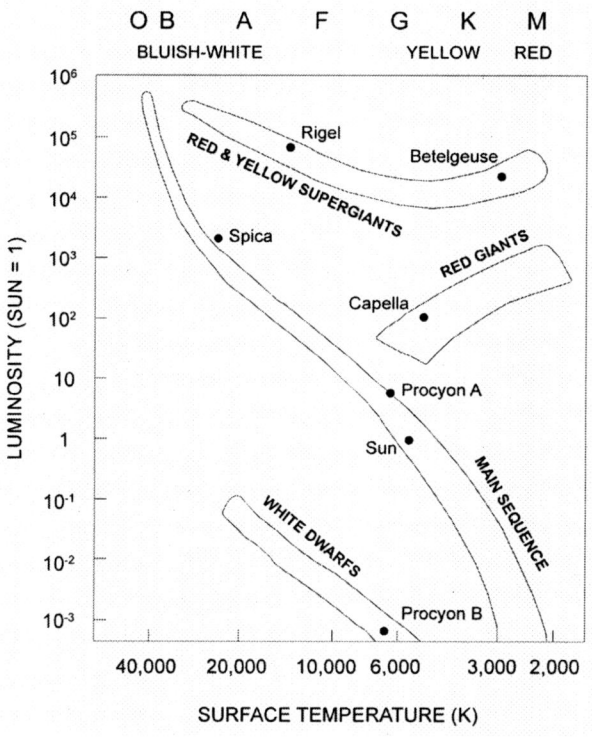

Fig. 5.7. The astronomers' famous HR Diagram.

spectral type is G2V; the letter G is in the figure, the number 2 is a subclass between G and K, while the Roman numeral V indicates that it is of relatively low luminosity within that class, such that our sun is a **dwarf**. A very large star in that class could be G2I, a **supergiant**. Pressures and temperatures are super high in the interiors of other supergiants, towards the upper left corner of the main sequence; their fusion process is aggressively fast and their lifetime short - as short as 10^7 years compared to 10^{10} for the sun. These are the young **O and B stars,** you will get to know them as they assume special significance when we consider all origins in later chapters;

astronomers also mention their "OandBstars" often.

Spectroscopists are the people who actually relate the temperature and output to the patterns they see in the lines of the spectrum caused by the absorption by various elements in the outer regions of the star. They assign to each pattern a **spectral type** with letters: O, B, A, F, G, K, or M, as we see in Fig. 5.7. We can do some of their systematizing, that the spectral type corresponds to a characteristic color: O and B stars are blue, F and G yellow, and those farther right and low in the diagram are red. The letters were first ordered alphabetically, A, B, ..., but later understanding changed the sequence so that the O – M sequence now appears rather arbitrary.

Table 5.1. Characteristics of a few Stars

Name of star	Constellation	Type of star	Distance (light-years)	Mass (M_{sun})	Radius (R_{sun})	Surface temperature (K)	Intrinsic brightness (L_{sun})
Procyon A	Canis Minor	Main sequence	11	1.5	2	6,500	6
Procyon B	Canis Minor	White dwarf	11	0.6	0.01	6,000	0.001
Capella	Auriga	Red giant	46	3.5	13	5,000	100
Betelgeuse	Orion	Red supergiant	600	20	500	3,000	20,000
Spica	Virgo	Main sequence	260	15	3	25,000	2.000
Rigel	Orion	Blue supergiant	800	20	36	15,000	60,000

Table 5.1 shows examples of characteristics for stars seen at various places in the HR diagram. Closest near us, see its distance in the last column, is a binary system of which one of its stars is not too different from the sun, Procyon A. Its companion is Procyon B, which is a **white dwarf**[4] that we shall discuss further in the following Section. I had remembered the Procyon name thinking it had something to do with proximity, but the dictionary has it after a Greek word for "fore-dog", in the constellation of Canis Minor (the small dog), rising before the brightest star in the sky, Sirius; the naming was done long before their distances were known.

One can well imagine from our discussion of constellations that a basic problem in astronomy is **to determine distances**. The stars in our skies show a variety in brightness, but that can be due either to difference in their output of radiation, or to difference in their distance from us. One of the methods to find distance is by inspection of the spectrum of the star and then in the HR-diagram finding from the ordinate the actual amount of the output of radiation for that spectrum (learning its actual brightness). Next, the astronomer measures at the telescope the apparent brightness as the star appears in the sky. Next, one compares the star's **actual** and **apparent** brightness to determine its distance (because the dimming depends on the distance just as the brightness of a candle depends on its distance). For a cluster or galaxy of stars, after having obtained the distance of only one star with that simple method, that is also the distance for the other stars. Then one can verify the output of radiation for various spectral types, etcetera; that is what spectroscopists do. There are complications that make the science interesting, such as a light-absorbing dust cloud on the way. It is interesting to apply those solutions, to learn how to assemble the universe in three dimensions. We must widen our view with a few special phenomena in the universe.

5.4 Supernovae and Black Holes

The O and B stars are so massive that their processes are accelerated gravitationally their "burning" in the transitions of hydrogen to helium is aggressive and their life time is short, only about ten million years. The star gets near the end of its life when the supply for the transitions runs out, with the first symptoms seen at its center where the fastest consumption occurs. The burning will then move outwards where some fresh material may still be available, in a thin shell some distance away from the center. However, the shift of heat generation away from the center causes a fragile configuration and vibration, the star is not stable anymore, the de-centralized configuration cannot stand the distortions caused by the instabilities; the star implodes and then explodes. It blows itself up and out in an enormously shattering event. This is a **supernova,** which appears as an exceptionally bright new star in the sky[5].

Astronomers and amateur observers search the skies to discover supernovae, and their reward comes with an international announcement of their success. The supernovae do not entirely disappear - some **end stage** is usually observed, as near the center of the spectacular "Crab Nebula" where the exploded star is known as a pulsar (see below). The explosions are not necessarily with the same force in all directions because internal motions and forces may be irregular, but they are always forceful, such that the end stage may be traveling, have a motion in space, because it had been propelled in that direction by the irregularity. The final stage in the death of a star generally depends on the mass of the original. The sun will end its life as a white dwarf. Stars that are originally more massive end as neutron stars or as black holes. We review the three in that order, as follows.

Once the core of a star like our sun, a star of **one** solar mass, fails to generate sufficient energy near its center, its outward pressures diminish steeply because these depend on radiation and gas compression. Its inward gravitation pressure is no longer balanced and sustained such that the interior becomes gradually unstable; it becomes a variable star of varying brightness, and the star eventually dies in a form we have already seen, that is called a white dwarf[4]. Here we need to discuss **density** as a basic concept for now and later. For a white dwarf such as Procyon B in Table 5.1, the density is as high as a million grams per cubic centimeter.

Let me give you that feel for the densities again by recalling first that water has 1 gram/cm^3, and we ourselves have 0.99 gram/cm^3 and so we can float in water. The above 10^6 gram/cm^3 therefore seems unimaginably high, but our own density might be seen as the one that is extreme in the stellar world, because it is due to **extreme spaciousness** of the whole atom that we will describe in detail in Sec. 8.3. We encounter high densities in the stars and much more so, in special objects; here come a few.

For stars that had an original mass greater than about **eight** solar masses, the inner core burns all its fuel to iron, which is a stable element, and eventually the atomic reactions stop and the outward support pressure decreases rapidly. The weight of the star then crashes the core so forcefully that it collapses into a non-space configuration of mainly neutrons (called neutrons because the positive protons and negative electrons have been crushed together to make a neutral particle). The outer layers fall onto the neutral and rather dead core, where they bounce off. The result is a giant supernova explosion[5], with an end state for the interior being a **neutron star**, while the remainder has been exploded away, which may be seen as a surrounding shell or nebula. The interior material had, in the meantime, collapsed from an original size of about 700,000 kilometers to only a few kilometers, with densities as high as 10^{15} gram/cm^3. Again, I want you to connect with that density not in a gee-whiz puzzlement or denial but rather recognizing it as a natural density because the protons and electrons had been crushed together by the high pressures into a small high-density star. Some of the neutron stars have a remarkable configuration, with only a small part of the star showing a beam of light, but that results in an excessive amount of radiation, due to pinching by magnetic fields. Because stars usually rotate, this object may at times beam its light like a lighthouse beacon towards us, or not, so that it appears to pulse on and off, and **pulsar** is therefore the name of this type of star.

Larger yet, original stars of some **twenty** solar masses, usually lose a lot of material during their evolution and then they end up as the above in supernovae, but in the case of original objects that consist of even more than twenty solar masses, it may be that they collapse to the configuration and phenomenon of a **black hole**[6]. The black hole forms either instantaneously, or first as a neutron star which accretes matter around it and all of it collapses into a black hole. In the process, a huge fireball may propagate its way out as a **gamma-ray burst** (Sec. 6.3).

The explanation of a black hole is elementary: the photon packages of radiation undergo such a strong gravitational pull towards the center that they cannot leave, they cannot escape. Another way to understand that intuitively is in terms of the speed needed for escape. For example, we cannot escape from the earth's gravitational pull because we cannot jump up with enough strength to go faster than the earth's **escape velocity** of 11.2 km/sec. The mass of a super-dense black hole is so large and its gravity so strong that the escape velocity is greater than the velocity of light, 300,000 km/sec. Light waves cannot move faster than the speed of light so they cannot escape from the hole, which is therefore black - hence the name of black hole.

We do have a few **observations of black holes** because some are bright in X-radiation caused by the violently collisional accretion of matter pulled away from a nearby star if there happens to be such a stellar companion. Anomalous effects in the light coming from that companion star are then observed. The overall explanation assumes an extremely massive hidden black hole pulling shards and flares of mass towards itself, and away from the companion star. Astrophysicists working in these fields have exciting times making discoveries with a variety of instrumental techniques on Earth and in space on objects in our galaxy and in other galaxies.

6. GALAXIES AND UNIVERSES

6.1 Interstellar Gas and Grains

After having looked at a variety of stellar objects, we will continue our voyage by considering their assembly into much larger configurations, namely the galaxies, and we begin with the overall appearance due to the stars in the Milky Way, and their background, the material between the stars. The observations ought to begin with field glasses or small telescope perhaps of a Planetarium, because the **Milky Way** (from the Greek word for milk, *galakt*) increase in brightness in that narrow strip of the sky is due to an increase in the number of **individual stars**, not a glow of gas or dust.

Next, when you are out on a clear and dark night away from city lights, you may see that there are areas where these stars seem to be missing. The best way to observe the sky above is by **lying on your back**, as is done by tourists for looking at the paintings on the ceiling of the Vatican's Sistine Chapel. You will then notice a first fascinating phenomenon: the Milky Way is irregular with dark patches! The density of stars is sharply diminished - it looks like there are dark clouds. One may wonder at first if the weather is cloudy - are we looking at clouds in our own sky? No, there is no motion - the next night the clouds are still at the same places!

What causes these clouds in the Milky Way? The answer is **interstellar grains,** alike the interplanetary ones of Sec. 2.1, small grains that scatter incident light into all directions. It is, in fact, difficult if not impossible to distinguish the two; at least some of the interplanetaries may well be original interstellar grains that drifted into the solar system.

When we are looking out from our solar system, stars may be behind clouds of those grains and a part or all of their light fades from our sight because it is being **scattering away**. The grains are small, but there are so many of them in those clouds that they can actually block the light. The grain is much too small for the naked eye to see if you had one in your hand to look at, too small to be resolved. Altogether, however, they can dim the light from a star appreciably because there are so very many of them on the long way from the distant star all the way to Earth.

Experts study the characteristics of the interstellar grains from their **interaction** with the starlight, from the scattering properties in different colors, at different wavelengths; classical studies have been made of such observations with special instruments at large telescopes. The interplanetary, or interstellar, grains are also collected with high-altitude aircraft and balloons and then available for exploration in the laboratory. The "Stardust" spacecraft flew through interplanetary space to collect them as well as the grains near Comet Wild 2 in 2004, it dropped them off in a capsule for a parachute landing on the Utah desert in 2006, and was a great success. Three other comets have been studied from spacecraft but this was the first sample return; a similar feat is being prepared for asteroidal material.

In addition to interstellar grains, there is **interstellar gas,** consisting of molecules drifting freely in space, not bound to grains. Images where starlight has been scattered and absorbed by the grains easily show glorious patterns of clouds of interstellar dust. Faint glows emitted by hydrogen molecules sometimes show clouds of interstellar gas. The reason for emphasis on the interstellar medium of gas and dust is that they play a primary role in the formation of the stars. Let us pursue this important point with professional observations of the composition of the material for such formation, made by specialized astronomers with special instrumentation on large telescopes and radio dishes. In imaging with sensitive electronic detectors, the dark areas in the Milky Way are observed in detail, and spectacular features are found that we like to give imaginative names such as the Horsehead Nebula and the North America Nebula.

There is also a discipline in discovery and identification of **molecules** in the interstellar gas. In order to identify another molecule, one computes its predicted lines in the spectrum of absorbed starlight with quantum physics. Sometimes it is possible to verify the computation in the laboratory, by observing the predicted lines in light that has passed through gas of that molecule. In any case, spectroscopic observations at an appropriate telescope allow careful comparison with the computed profile of many lines, and if there is a match, another discovery of a molecule in space is celebrated. The astronomer goes back to make the same observations in several directions and distances, not just to verify her discovery, but also to establish that the spectrum is the same everywhere, and that is essential for this book, namely to establish that there is **no variation in inorganic**

evolution, that the molecules are always exactly the same.

Table 6.1 shows a selection of 4 from about **150 interstellar molecules** that have been discovered (2007). The very first discoveries in

Table 6.1. Examples of Interstellar Molecules

Molecule	Symbol	First	Notes
Hydrogen	H_2	1970	UV, 1013-1108 A, NRL rocket
Ethanol	CH_3CH_2OH	1974	Radio, 2.9-3.5mm NRAO 11-m
Methane	CH_4	1977	Infrared, KPNO 4-m
Glycine	NH_2CH_2COOH	2003	NRAO 12-m, still disputed, for lack of laboratory calibration.

the 1930s were tentative, but by the 1970s the field and identifications became precise and professional, as the Table indicates. In 1970, a rocket of the Naval Research Laboratory allowed the identification of molecular hydrogen in the extreme ultraviolet above the earth's atmosphere. About one-third of the 150 molecules came from the 11-meter dish of the National Radio Astronomical Observatory (NRAO), on **Kitt Peak**, Arizona. The discovery of methane, in the infrared part of the spectrum, was made at he 4-meter telescope of the Kitt Peak National Observatory (KPNO). The 11-m telescope was upgraded, is now known as the 12-meter, and there was a change in administration such that it is called the Arizona Radio Observatory. A first discovery of an organic molecule, Glycine, was exciting, but required further study. Other highly complex molecules are being debated[1]. The general impression one gets from Table 6.1 is of the complexity of the molecules in space. Here is the factory of our molecules.

6.2 Our Milky Way Galaxy

The next observation to note is how **narrow** the Milky Way is. If that glow is the cross-section with our whole galaxy, it must mean that the stars and clouds are concentrated towards a thin and flat galactic plane, the central plane of our galaxy (Fig. 6.1). We have encountered another configuration of a narrow plane before, but at a far smaller scale, the ecliptic, when we saw that most of the planets orbit quite close to the plane of the ecliptic. The similarity does indeed indicate that there is a common mechanism of formation for galaxies and planetary systems, a mechanism that causes **concentration of bodies towards a plane**. We should note and remember for now that the flattening is of the greatest importance for understanding the details of origins and evolution that we will study in Chapters 9 and 10.

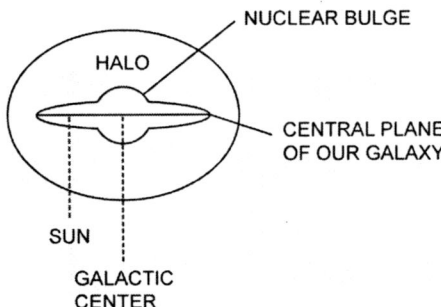

Fig. 6.1. A sketch of a side view of our
entire galaxy, for baryonic "visible"
matter, not including dark matter.

The next observation is for a confirmation that the Milky Way seems to divide **the sky in equal halves**. If it were not true, the "dark sky" off one side of the Milky Way would not be as dark as the other side because more of the galaxy would show on that brighter side. The two dark sides are nearly equally dark so that the solar system must be located rather close to the middle within the plane of the galaxy, and that position is indicated by **CENTRAL PLANE OF OUR GALAXY** in Fig. 6.1 and by the dark horizontal line in Fig. 6.2. Actually, if you have good eyes you can see that the dark sky is not quite as dark on one side as on the other, so it follows that the solar system does not lie exactly in the central plane.

The Milky Way band encircles the sky with the stars of the galaxy concentrated in the shape of a discus and we are deep inside it. In the left-right directions of Fig. 6.2, the location of the sun within the central plane is way off from the center of the galaxy. Our solar system lies about halfway from the center to the outer rim as indicated in Fig. 6.2 and it follows that we should see more stars in the direction towards the center of the Galaxy than in the opposite direction. Yes, naked-eye observations confirm that. The Milky Way is seen unevenly in our skies: the part we see in May - July looks wide and dense because we are looking towards the **central bulge**. From the Northern Hemisphere, the bulging is in the Southern parts of the sky; it is therefore more glorious when seen from the Southern Hemisphere, where it stands overhead and is broad and richly crowded with stars and nebulae (see Fig. 6.7, below). On the other hand, the part of the Milky Way opposite to the bright center is much sparser because we are looking away from the central bulge; this is seen best from the Northern Hemisphere where it stands high in the sky from November to January, and it seems thin and unpopulated in comparison.

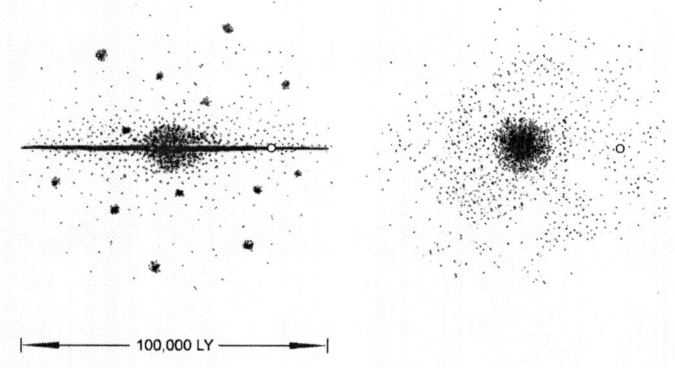

Fig. 6.2. Edge-on and pole-on views of our galaxy, with the sun now sketched on the right (open circle).

We have derived important and a rather intricate conclusions from visual observations. The detailed studies of our galaxy require a rich variety of **observational facilities** such as infrared instruments and radio telescopes, along with **theoretical studies** for the formation of star systems and for the dynamics of moving objects. Our galaxy has about 10^{11} stars and a great variety of clouds consisting of gas and dust, and it all moves through endless turbulence and dynamics.

How large is our Galaxy? At the speed of light, it takes about 26,000 years to travel from the center of our galaxy to reach us[2]. We say then that the sun is located at **26,000 lightyears** from the center. We are roughly 24,000 lightyears from the outermost visible stars in the other direction (the right end of Fig. 6.2); this is approximate of course because there is no sharp edge to the Galaxy. The numbers show that we are located a little more than halfway from the center to the edge of our galaxy. It is simple to memorize the sum of these two numbers as an even 50,000 and that the diameter of the galaxy therefore is about **100,000 (10^5) lightyears**.

For the study of our universe we are going to look in directions away from the Milky Way [up and down directions in Figs. 6.1 and 6.2 (left)]. Here we can look out because we will see fewer stars that belong to our galaxy, and the skies are therefore **dark** for penetrating **deep** into intergalactic space. In our imagination we travel in the direction perpendicular to the plane of the Milky Way, for example up in Fig. 6.1, and stop at a considerable distance, to look back at our galaxy from that vantage point. Indeed, if our solar system had been located away from the Milky Way (instead of being close to the middle within its plane) you can imagine that our galaxy would have appeared as a large object in the sky. Our view of other galaxies in the downward direction would be sparse due to too much of our own galaxy being in the way. We are fortunate to be close to the Central Plane, but not close to the big and crowded center of our galaxy.

Fig. 6.3. A spiral galaxy seen edge-on. The dusty central plane is warped, probably as a result of an encounter with another galaxy.

Observational cosmologists are applying for much time at the largest of telescopes and use it for making observations in these directions away from our Milky Way, in order to study other galaxies in large variety (a special case is in Fig. 6.3). Here we need introductory remarks about our own galaxy for the comparison with the other galaxies. For that purpose, please imagine again that you would be still up there at that vantage point and looking back from a considerable distance. I will tell you what you would actually see in that large object, namely a beautiful configuration, with the most striking feature that the dust and gas clouds and the young stars and open clusters are concentrated in spiraling configurations.

We speak of **spiral arms** and a **spiral galaxy** as seen in Fig. 6.4. The spiral structure is not intuitively obvious to understand, but computer simulations have reproduced it for objects in collision, simulating a system that is in rotation, which is typical in the universe (see Fig. 9.1, below), and that is important to take into account[2]. The essential component of the simulations is that rotation, about the galactic center, a fascinating effect studied in detail in order to comprehend the dynamics of the ensemble and of the spiral arms.

OUR MILKY WAY GALAXY

Fig. 6.4. A regular spiral galaxy. This is the Andromeda Nebula, M31.

At the distance of 26,000 lightyears, the solar system orbits about our galaxy's center with a velocity of half a million kilometers per hour! Yes, we surely are moving in space and in a variety of ways for us to jubilate about:

- centimeters per year by plate tectonics;
- a quarter of a meter up and down per day by the tides;
- nearly 1,700 km/hr at the equator by the earth's rotation;
- 18,000 km/hr by other asteroids for a future astronaut on one in the main belt;
- 108,000 km/hr in the earth's orbit about the sun;
- 500,000 km/hr rotation about the center of our galaxy;
- and our galaxy moves at a million km/hr or so with respect to other galaxies.

Back up to our vantage point at some distance from our galaxy, the next we would probably notice are the globular clusters at great distances from the galactic plane. They occur mostly in the halo, while open clusters are located mostly in the galactic plane where there is a lot of interstellar gas and dust. We will use these observations in Chapter 9 and it will then be clear that the **open clusters are young** and still concentrated near the plane of the galaxy where they were born. In contrast, **globular clusters are older** and more the result of collisions and violence that happened at various times in their history.

The globular clusters have an up-and-down motion in Fig. 6.1 and in the left hand sketch of Fig. 6.2. The mass of the galaxy is mostly concentrated near the central plane and gravity attracts the globular clusters towards the plane whenever they are above or below it. They then move towards the plane and most likely without retardation in collision with material there because there is much space in-between, so they shoot through to the other side. There the same action occurs, of the gravity slowing them down, and pulling them back again. From this observed motion of globular clusters, **an amazing discovery** followed, already in 1905. The attracting gravity and mass in the galactic plane were computed from these motions and it was found that there is more mass attracting the clusters than we are familiar with from all the objects that we can see; this became first known as the **missing mass** problem, now it is called **dark matter**. Also from studies of galactic rotation, and later in several other ways, the puzzling conclusion came that only about **4% of the universe** is in matter we know of, such as atoms, molecules, people, planets, stars, and nebulae. Only about 4% of the universe is visible – we do not see 96% of the mass of the universe. This topic is in the forefront of studies with large telescopes and expensive spacecraft, and we learned that the 96% is divided in 22% **dark matter** in galaxies, and 74% in some form of **dark energy** of expanding intergalactic space (the three percentages and the expansion are explained in Sec. 6.5). The 4% may help us not to think overly much of ourselves in the universe, not to have anthropic notions or principles.

Fig. 6.5. A galaxy presumably like ours, having a barred structure at its center.

Still at our vantage point way up there, we would see again that **the universe is mostly space**, even when there are many objects in the plane of the galaxy. Our classic Fig. 4.7 of the Oort Cloud gives a good impression of the spaciousness, how the typical distance between stars located near the plane of the Milky Way galaxy is several lightyears (and much greater away from the galactic plane). We would see that the interstellar clouds also appear to be rather isolated, occupying only a few percent of the space between the stars.

We would surely notice that the center of the galaxy is a dense region of space. From the earth, it is difficult to penetrate, because we too are close to the galactic plane; the obscuring interstellar dust adds up all the way in, and increases towards the center. Some penetration around the grains succeeds with long-wave infrared wavelengths, while X-rays and gamma rays are energetic enough to penetrate through the grains. The center of the galaxy appears to have a source of **great power**, an enormous and violently active black hole perhaps, or several of them. The sum of that concentration adds up to several million times the sun, but we do not know too well yet what types of energetic objects are present. Also rather hidden is a cylindrical structure at its center; ours is therefore called a **barred spiral galaxy** (Fig. 6.5), and that bar structure is not yet understood.

Let us look at other galaxies and at the most energetic objects in the universe.

6.3 Other Galaxies and Energetic Objects

There are three types of galaxies, probably occurring in this order of evolution: the earliest few percent are **irregular**, next in time are **spiral** galaxies, and the most frequent and old type has a more **elliptical** shape.

Fig. 6.6. An irregular galaxy.

Edwin Hubble had the opposite order, but the present one comes from direct observation with the largest of modern telescopes, and these observations are made in carefully selected places where there are not too many stars in the foreground, in our own galaxy. The surprising result was that the irregulars occur at the greatest distance, the longest time ago – they are the originals.

Irregular is **a catch name** for all sorts of shapes, without orderly structure (Fig. 6.6). Spirals have the dust and stars that we saw in the spiral arms and they are in the process of making stars. Elliptical galaxies seem to have few interstellar dust clouds, there is not so much material left for making new stars, and the stars that are observed are determined, from their spectra, to be on average quite old.

A special case of two irregular galaxies are the Large and Small **Magellanic Clouds,** seen with the naked eye as two small areas of hazy light, rather widely separated in the skies of the southern hemisphere. The names are after the famous explorer Fernão de Magalhães (1480-1522). Figure 6.7 shows them in the magnificent southern skies the way he saw

Fig. 6.7. View of the center of the Milky Way, the Magellanic Clouds, and of two telescope domes at the Cerro Tololo International Observatory in Chile. The only source of illumination is starlight.

them, but he could of course not have dreamed of such large telescopes there; in fact, there were no telescopes yet in his day.

The identification of individual galaxies relies on catalogues in which they are collected, such as the one made by Messier, a French astronomer. An example of a special case is Messier 31, **M31**, because it is the farthest object that our eyes can see, albeit merely as a small hazy patch. Because it is so near, bright, and extended, it is imaged best with a small telescope as was done for Fig. 6.4. It is a spiral galaxy, somewhat like ours but for the bar at the center of ours (Fig. 6.5). **Andromeda Nebula** is the more common name as it occurs in that constellation.

Spiral galaxies generally have nuclei with enormous energy generation as we saw for our galaxy. The study of **Active Galactic Nuclei** is in the forefront of astrophysical research; the power source may be a black hole (see below), or several black holes, for a total of $10^6 - 10^9$ solar masses and even then accreting additional matter.

Observations and interpretations of such massive giants use **relativity theory**[3]; there is a well-known example of what it can do that has been used since the 1920s. I am referring to the bending of light rays when they graze a massive gravitational body like the sun. Einstein predicted their **bending** in his relativity theory, but the verification required careful effort to observe during a solar eclipse. During totality, the sky is dark enough for a few minutes to see and image stars in the background. One does the same imaging again after the sun has moved away from that star field. The one with the eclipsed sun will show the stars in slightly unusual positions because the sun's gravity had bent the rays.

There is now similar but more advanced imaging of the bending around massive objects, on a galaxy remotely in the far distance, while it so happens that there is another galaxy in the foreground. That is not a rare event because there are so many galaxies seen in deep space. Anyway, the light beams from the distant one are bent by the mass of the foreground galaxy on its way to us. Two small images of the distant galaxy may be seen to two sides of the galaxy that is in the foreground; if the obscured source is precisely centered behind the galaxy in the foreground, a set of quite small images is seen all around the latter. The experts call this by the appropriate name of **gravitational lensing** while the observations provide understanding of galaxies and relativity theory[3].

Gamma ray bursters (GRB) are in a very different category; they show during a few seconds or minutes and then are gone; they do not show repeatedly from the same direction. Their distance in space has been determined for a few events at billions of lightyears. Being seen, even though they are at large distance, implies that they are highly energetic; they seem to put more energy out in a short burst than what the sun emits into all directions in a billion years. These are indeed the most powerful explosions in the universe. They too may be beaming only into a small part of the sky as we reported for the pulsars. The GRBs are still mysterious. One of the hypotheses is that they come from the collapse of a star having more than 30 times the mass of the sun so that its core collapses into a black hole. The tremendous energy of the collapse and the in-fall of matter into and onto the black holes might then power the gamma-ray burst.

There are two kinds of **cosmic ray** energetic particles bombarding the earth. First, there are the ones that come from supernova explosions or from the time of origin of the universe. The other cosmic rays are **solar energetic particles,** which are accelerated by magnetic effects in the loops of solar flares (see one in Fig. 5.1). In either case however, the cosmic rays do not directly penetrate the terrestrial atmosphere; the particles that are zipping around at the surface of the earth are **secondaries** produced by interaction of the cosmic rays with nuclei of atoms in the upper atmosphere. In any case, to speak of cosmic "rays" is misleading since particles are involved, mostly protons and electrons, different from the gamma rays for instance, which are photons of electromagnetic radiation such as visible light for instance. These various energetic objects serve especially for observing distant parts of the universe.

6.4 Observing the Universe

We shall now complete our voyage outward to the limits of the universe, millions, and then billions of lightyears away. The total number of stars is about 10^{21} in terms of solar masses, which is easy to remember from the 10^{11} stars in our galaxy and about 10^{10} galaxies in the universe. For a universe however, we shall no longer consider the stars individually nor the interstellar medium in galaxies, but switch to considering whole galaxies and what lies in between them. One of the frontier topics of astronomical research concerns the **inter*galactic* medium (IGM).** This is in the space between the galaxies, it is not empty; here lies a field for modern research. Debris from the aging of atomic particles and aging stars such as white dwarfs may be leaking out beyond the gravity of their galaxies. Active Galactic Nuclei in clusters of galaxies may account for most of the radiation between galaxies, but there still seem to be unexplained source(s). An unknown example is the general X-ray background radiation discovered in early rocket experiments.

Galaxies occur in groupings. Outside of our Milky Way galaxy, we first encounter the others of our **Local Group**, which has about 30 members and is about three million lightyears in size. Such local groups are members of **clusters**, which have a million or so galaxies, and **superclusters,** which have some ten thousand clusters. The National Geographic Atlas of the World[4] has illustrations for studying the three stages. An example of a cluster is in the constellation Virgo, the Virgo Cluster, another example is the Coma Cluster.

Here is a fabulous thought experiment. Einstein had found from his math and common sense that galaxies could not just be hanging out there motionless, but had to be moving in their mutual gravity fields. Imagine that you are a galaxy in space, rather lonesome because the void is so large and dark, but still, your gravity will make you move towards whichever galaxy wins over the gravitational tugging by all the others because it may be closer or more massive. In fact, the Andromeda Galaxy is moving towards ours. This helps us to think in a broader sense of the **interaction of everything** with everything else. This applies on small scales in our environment, in our universe with galaxies causing motion, groupings, clusters, and collisions[5], and we will see the same for universes in Sec. 6.6.

The study of faint objects is possible through observations with the largest of new telescopes on sites where the air is steadiest for pinpointing the starlight sharply with the least of turbulence. The pinpointing makes sharp images of the light concentrated to make them the brightest, which is essential for observing the faintest of objects. **Chile** and the big island of **Hawaii** have the best sites, with telescopes having mirrors as large as 10 meters in diameter (diameters are used not just because they sound larger than radii, but also because the dome and its expenses are affected by that larger number). These are the days when large telescopes of all kinds are proposed and built for the study of the most distant objects, when they were young. A telescope with two 8.4-meter mirrors on the same mounting, the Large Binocular Telescope, is on Mount Graham east of Tucson, a proud feat of the University of Arizona's Astronomy Department. Special feats were also to build the Arcminute Cosmology Bolometer Array Receiver (ACBAR) and other telescopes in Antarctica, which has exceptional conditions of steady atmosphere. An international but mostly European consortium is developing a site at 4,900 m elevation in the northern deserts of Chile - where there hardly ever are clouds and rain - for a 30-m optical telescope as well as a large radio telescope. The second highest elevation is of the Indian National Observatory at 4,500 m, East of Leh in the western Himalaya; astronomers in the south of India control the operation and data analysis via an Indian satellite. This is a proud feat for India, which is still sometimes called a developing nation.

The least atmospheric turbulence and absorption occur of course outside of the atmosphere, **in space**, with the additional advantage of observing at wavelengths that are inaccessible on Earth (Fig. 1.8). Regarding the turbulence we should, however, note that modern earth-based telescopes are equipped with equipment to compensate for image motion, **adaptive optics**. For the studies at a variety of wavelengths, a wide range of space telescopes is needed (Sec. 1.4). The 2.4-m Hubble Space Telescope is an example, but it has a duty cycle of only about 15%, largely because it is in earth orbit of about 90 minutes, causing interruptions in long exposures when the Earth is in the way of observing the object. About half of the radiation of the cosmos is at long wavelengths, in the infrared, as is emphasized for the Spitzer and Herschel telescopes in space. Next is planned the 6.5-meter James Webb Space Telescope for launch in 2013, and parked at a much greater distance from the earth for greater efficiency. It is also to observe in the infrared part of the spectrum only, especially aimed at studying galaxies when they were young, between one million and a few billion years old.

Here are some spacecraft for galactic and extra-galactic studies, and a showing how the names have great variety. The HST was in honor of astronomer Edwin Powell Hubble, who was one of the pioneers in cosmology working with telescopes of 2.5 and 5-meters in diameter on mountains in southern California, Lyman Spitzer was one of the early pioneers for astronomical telescopes in space, and James Webb was in charge of NASA in its early days. The German ROSAT (which ended in 1999) had been named for physicist Wilhelm Konrad Röntgen who was the discoverer of X-ray techniques in the medical world - Europeans speak of **Röntgen radiation** instead of X-rays. The Newton is after Isaac Newton, of course, the Rossi X-ray Timing Explorer (RXTE) after physicist Bruno Rossi, "Chandra" was accepted after the suggestion by a grade-school student for naming after astrophysicist Chandrasekhar, and Beppo-SAX after physicist Giuseppe (Beppo) Occhialini. SAX stands for Satellite per Astronomia in Raggi-X. The European INTEGRAL mission observes gamma-rays, black holes and other exotic objects over a wide range of wavelengths. NASA has a spacecraft to study gamma-ray bursters, called Swift; that name refers to the satellite's ability for swiftly pointing on-board telescopes, and several Earth-based ones as well, to a newly discovered burst and then studying its afterglow for detailed observation[6]. The "Laser Interferometric Gravitational Wave Observatory" is already in operation, as is the "Gravity Probe B, The Relativity Mission". The "Laser Interferometer Space Antenna" is to observe the merging of black holes, "Constellation X Observatory" will aim at gas that gets too close to a black hole, and there will be a "Black-Hole Finder Probe". Instrumentation that is specifically involved with what happened at 400,000 years of age of the universe will be at the end of Sec. 7.5. one of these is "The Joint Dark-Energy Mission" to study the universe's expansion, as we shall do next.

6.5 Expansion of Intergalactic Space

The discovery that intergalactic space is expanding was an unexpected finding, made by a rather unknown astronomer at a small observatory, and no one understood it for years. Please recall from Sec. 6.2 how spectra show sharp lines in well-known patterns that are due to absorption of the light by hydrogen and other elements in stellar atmospheres. It was astronomer Vesto M. Slipher (1875–1969) at the Lowell Observatory in Flagstaff, Arizona, who had noticed already in 1912 that the patterns of absorption lines in some **nebulae** were not in their usual place. He pursued the topic for a dozen years observing some 40 nebulae, and found for most that the lines had shifted to longer wavelengths. He already knew from studies of stars that the shifts of the patterns reflect motions towards or away from us. The shifts make sense on the assumption that the waves stretch by motion away from us; when their wavelengths become longer, they have a **redshift**[7]. The case of the stars was already well understood, and some of them showed waves pushed together by the fast motion towards us, thus the wavelengths are "blue shifted".

This type of effect is similar to the Doppler Effect for sound, after Austrian physicist Christian Doppler, which is for motion **through** space, while here we are dealing with stretching of wavelengths **by the expansion of space**; furthermore, relativistic effects enter in at high velocity, near the speed of light. Anyway, Doppler explained that people in the days of steam trains were hearing a high pitch, of waves pushed together, when the whistle-blowing train approached, and a low pitch, of stretched waves, after the train had passed and was speeding away[7]. He demonstrated the effect by placing musicians on a flatbed railroad car, all blowing the same note precisely. I tried it with our car horn screaming along while racing at top speed by our two little sons and their mother standing on the side of a straight and deserted desert road, a frightening experiment. They did not hear any higher and then lower notes. Pop's experiment had failed!

It took me some time to restore my reputation by figuring out that the

Doppler Effect did not work because the car-horn's wavelengths occur over a broad range, not sharply defined, the note or **pitch** is not narrow enough so that the effect does not stand out from the width of the horn's wide spectrum. This is also sadly the case for modern train horns, so we do not hear this delightful effect along the railroad tracks anymore either.

In astronomy, a debate occurred over the realm of the nebulae that Slipher and later others were observing. Edwin Hubble had joined the field and he became famous because he had access to powerful telescopes in Southern California. The essential problem was the sorting out of so many **fuzzy patches in the sky, the nebulae**, which was a great puzzle at that time. Where were they, were they inside or outside of our galaxy? At what distances?

Thanks to pioneering by Slipher and Hubble, and the work of many other observers, we now understand how many different types of galactic and extra-galactic nebulae there actually are. Some are regions of star formation as the Orion Nebula is, relatively nearby and well within our own galaxy. Others can be distant, **extra-galactic**, and yet look the same at least in the older smaller telescopes. Their distances are determined with special methods; such determination is a major problem in astronomy; we described the basic method for stars in the HR diagram at the end of Sec. 5.3.

All but one of the galaxies show the spectral lines shifted red-ward. M31, the Andromeda Nebula, is a spiral galaxy like ours, at only two million lightyears' distance, and is the striking exception with a blue-shift, as it moves towards us because of the mutual gravitational attraction of the two galaxies (the encounter will be described in Sec. 12.1).

All other galaxies are moving away, and the phenomenon is the expansion of **intergalactic space**; the stretching of the wavelengths occurs because space is expanding. The space inside individual galaxies, or within us, is not expanding, because the components of galaxies, stars, atoms, and bodies have their own gravity regimes. The expansion of extragalactic space occurs in all directions as is known from observations of various galaxies in various directions and at various distances. When at the greater distance more space expands, the spectral lines are shifted more and we measure greater speed. A loaf of raisin bread rising in the oven is doing the same. Just imagine that you are one of the raisins in the dough. You see a fellow raisin in the distance moving away from you as the dough rises. Now look at another one in the same direction but twice as far away; this one moves twice as fast because there is twice as much expanding material in between.

To generalize our understanding we must summarize the observations of the rate of expansion, we must make a **law for the redshift**. Stars and galaxies are, of course, faint at great distance (just as headlights of distant cars are faint) so that large telescopes are used with special techniques to determine the spectra. This work has in fact become a frontier of cosmology - observing ever fainter and farther with sophisticated instrumentation. Instead of distance determinations using stars (Sec. 5.3), which now are too faint, the apparent brightness of active galactic nuclei is used. We shall remember the summary of the redshift observations as follows. **An object receding near the velocity of light is at about 1.373×10^{10} lightyears distance.** One can use the rule to compute, linearly, the distances for other expansion velocities and *vice versa*.

The basic assumption held for a long time was that the expansion is indeed linear, proportional with time, but this seems no longer true. There are now observations indicating that since the age of the universe of about 6×10^9 years, the expansion may have changed to slightly **accelerating**[7]. Theorists are having a challenging time trying to explain what might have caused the reversal; ideas like anti-gravity forces or leakage of gravity are doing the rounds[7], while we will have a simple explanation, below. Einstein had discussed some of these possibilities when he found from his math and common sense that galaxies could not just be hanging out there motionless, but had to be moving in their mutual gravity fields. The universe had to be collapsing or expanding. It could be a pulsating universe, slowing now, stopping, collapsing, and somehow reversing into an expansion again (Sec. 7.2). That was around 1916, and the astronomers believed from limited data that there was no motion, so Einstein added an artificial term in his equations, a **"cosmological constant"**, to make them agree with the observations. About the year 1926, however, Hubble showed convincingly that there was expansion, and Einstein removed the term. The term is now used in detailed theory.

The non-linearity is small, so we continue to use the assumption of linearity for the red shift. As one observes at ever-greater distances with large telescopes, one should encounter a distance at which the expansion moves at the **ultimate** rate, at the velocity of light. Indeed, the largest red shifts indicate that their sources are receding close to the speed of light. Clearly, we cannot observe something moving away from us at speeds greater than that, because the light it emits could never reach us. The limit of such an observable universe, beyond which we can no longer observe anything, is therefore the distance away from us where an object would be receding at the velocity of light. Using the above boldface summary that is 1.373×10^{10} lightyears, 13.73 billion as one says in the U.S. The fundamental and presently unsolved question is whether this is the observable limit, or a real edge of our universe.

We should also consider the **age** of the universe. If we assume that it has always been expanding steadily, if there had been no acceleration, the question is how long the expansion has been going on, what is the age of the universe?. The age and other parameters are actually determined from detailed analyses made by various groups of investigators, and the answer is **1.373 x 10^{10} years**.

In preparation for the discussion of other universes in the next Section, we shall now summarize whatever little we know at this time about the 4%, 22%, and 74% compositions of our universe (Sec. 6.2). The first, 4%, is everything we can observe, from ourselves to society, from dust to meteorites, from comets to planets, and from stars to the farthest galaxies; a technical name for it is the baryon universe, where baryons are particles like the protons. The 22% of **dark matter** occurs anywhere in and around galaxies, has been known for years as the missing mass, it is extensively observed, but it still is not physically known. The greatest puzzle is that our galaxy has increasing dark matter outwards, much larger than in Fig. 6.1, which is for entire baryon galaxy only. The 74% in form of **dark energy** is an actively studied topic, as is dark matter, but even more of a mystery. Other identifications for it are anti-gravity, and the cosmological constant; in any case it is connected with the acceleration of the expansion of intergalactic space and it may thereby control the shape and spacing of galaxies[7]. These concepts lead us to even greater distances, of other universes.

6.6 The Multiverse

Having come with our descriptions to the end of our universe as far as we can observe it, we ask the question, "Why would our universe be the only one?" There is no direct observational evidence of other universes as yet, but theoreticians have models regarding the existence of other universes[8].

Some of the books written are based on **anthropic notions** such as that we are so special, so rare, that there must also be many other universes without people. Astrophysicist Fred Hoyle pointed out, already decades ago, that it is extremely unlikely to have the **fine tuning** that is observed for the nuclear transitions of Fig. 5.4 inside O and B stars for producing such complicated stars, let alone the resultant complexities of humanity[8]. If the physical constants of the elements were even slightly different, the selections and combinations could not occur. Therefore, some authors of books believe that many universes originated such that one of them could be ours, just by chance. We shall stay with the more scientific literature.

Physicist Lee Smolin[9] developed a model of producing universes abundantly from black holes. The word "multiverse" was probably used first by him, for all universes, and we will stay with that definition. Astrophysicist Martin Rees[8] critiqued Smolin's work, and made his own model for various universes having various physical constants and conditions.

Dozens of physicists work on **theories of inflation** in which they rely on quantum mechanics that space is not an empty void but a space-time background that through quantum fluctuations is able to produce our universe[10]. Inflation is then a rapid expansion away from that background, with thorough interaction of components, which must have occurred in order to explain the uniformity of our universe that is observed with spacecraft and Earth-based telescopes (Sec. 7.5). Inflation theories are well-established through the international participation of scientists, while data observed with spacecraft begin to provide confirmation. Since the 1990s, the models derive that the space-time background is continuously spawning new universes from that quantum-mechanical background (space-time is the unification of space and time according to Einstein's theories). To me it seems unlikely that masses of universes, some 10^{23} stellar masses each, can spring from a vacuum. I have not seen this discussed, while the models do speak of a multiverse with an abundance in variety of universes[10]. The inflation theories may not be needed anymore if the $M(\alpha)$ model becomes accepted and established.

In summary, there are research and philosophy regarding other universes. I end this section with that $M(\alpha)$ model on the multiverse I have been working on for years. It uses a mathematical expression predicting **masses and physics of original stars, universes, and the multiverse**[11]. Two papers were published in a new electronic manner of NASA and the Los Alamos National Laboratory, which accepts authors on the basis of their scientific record, and individual papers are not refereed, which expedites the publication of totally new approaches; in my case, it is all new. Should the model be proven wrong, that will affect only what I write about my own conclusions regarding the multiverse.

An expression had been derived in the 1930s by my Professor S. Chandrasekhar, and I use it in a simplified form, $M(\alpha) = (hc/G)^\alpha$ where $M(\alpha)$ is the mass of original cosmic objects; h is Planck's constant, c the velocity of light, and G is Newton's gravitational constant. The masses of $M(\alpha)$ must however be expressed in terms of the proton mass H, which is a universal constant, for all societies in the multiverse to use, rather than our anthropic kilograms or solar masses. The **power of (hc/G)** is that it shows with h the presence of quantum physics, c for relativity, G for gravity, while H represents particle physics; these are the basic physics we depend on and understand rather well, and the four are here unified in their application. Chandrasekhar had derived the structure, composition and source of energy of stars with related equations like that. For original stars the exponent $\alpha = 1.50$ yields a good fit, and for $\alpha = 2.00$ it is the mass of our universe.

Here we concentrate on values of $\alpha = 2.50$ and larger, which are the steps beyond our own universe, outside of it. The following chapter stresses a basic characteristic of evolution, that when any option such as $\alpha = 2.50$ appears, it is filled. There is no reason why nature would not use it, or why we could not use it to study the multiverse; there is no restriction here by the velocity of light that we saw in our universe near the highest redshifts in Sec. 6.5. Values of $M(\alpha)$ are indeed open for any value of α, all the way into infinity. Section 8.2 shows how our world is quantized, and here the quantization uses steps in α of 0.50, from 2.00 for our universe, to 2.50 for a cluster of universes, to 3.00 to a supercluster, etc. The result is as many as 10^{19} universes at $\alpha = 2.50$ (which is the same step as for 10^{19} original stars in our universe), 10^{38} universes at $\alpha = 3.00$, and so forth, a tremendous multiverse of universes.

The first of the conclusions concerns the **critical mass**, which is defined as not too large, or the universe would gravitationally collapse under its own weight, and not too small, or shortage of gravitational pull would allow the universe to expand apart into nothingness (not enough gravity to hold it together). The above mass for $\alpha = 2.00$ is indeed near that critical mass for our universe and for all others in the multiverse. It is seen that near-critical mass is in fact a **condition** for any model of multiverses; their universes must all have that same mass, or they will not be stable and the multiverse itself would not be enduring. One can imagine evolution to take place in the multiverse with any universe of mass different from near-critical mass simply not surviving.

Actually, the mass is a little less than exactly critical, it is a factor 1.94 smaller than critical, and that probably evolved so that all universes expand fast enough for necessary interaction. The effect is seen from above a quiet pond on which raindrops fall. The rings on the water expand and travel through each other. The result in three dimensions is a thorough mixing of the debris from old galaxies that we shall discuss below. The universes will then be similar in composition because the mixing brings material together from many other old universes. Furthermore, since (hc/G) and expression in terms of proton mass are used, every one of these universes has the same quantum, relativity, gravity, and particle physics represented by h, c, G, and H. All universes are therefore **the same in physics and mass.** They could therefore produce life as ours does.

The second major conclusion follows, that the **origin of evolution** together with all basics and natural laws occurred in the multiverse, where they always search in trial-and-error for survival, finding the 1.94-critical mass and these physics. Any universe outside of these characteristics does not survive; Sec. 7.3 works this out in further detail. In the meantime, Hoyle's problem is solved, because his fine tuning developed already in the multiverse with self-tuned mass and physics for each of the universes, and that is how it was done for our universe as well.

The model stimulates a new discipline of studies, of which a few examples are mentioned here. We have already noted how **galaxies are aging** with the gradual transition from the original hydrogen and helium to heavier elements, shown in Fig. 5.4. Even the strongest nucleus, the proton, eventually breaks apart - Andrei Sakharov studied that - and the larger nuclei have weaker bonds so that they age even faster. The old galaxies are floating, within their own universes and the multiverse, spreading their debris of sub-atomic particles, radiation (photons), dark and energetic matter and whatever else is left. By the expansion of their intergalactic space, the old galaxies float through their own and all other universes, such that they encounter and share and mix their material.

The M(α) model considers the **demand** of newly spawned universes to be in balance with the **supply** of mass and energy from decaying universes. The debris from decaying galaxies ripples in all directions on the expansions of their intergalactic space. It probably consists of the broken up components of protons and other atomic nuclei, these are the quarks and other subatomic particles, their temperature is low in space, near 0 K. There also may be photons from old but originally bright sources such as active galactic nuclei (AGNs); their aging is a red-shifting to temperatures also close to absolute zero.

Long time spans are involved in all these processes and particularly for the material from various galaxies to stick and grow together as we saw happening in the interstellar medium. Eventually, a huge mass of debris may be accreted by its own increasing gravity. Material to build a new universe is present, as well as the energy to put the subatomic particles together again into atomic particles; that energy is provided by the gravity of the accreted mass. The recombination towards complete protons can probably begin wherever the density of space in the accreted mass reaches the particle density of protons, which is about 10^{18} kilograms per cubic meter. That would be near the center of the mass first. There is **no bang**, there is merely a gentle new start of a universe.

The model is complete in outline, but new studies need to develop it more in detail. The conclusions follow from straight application of the $M(\alpha)$ equation, without artificial or anthropic assumptions, and yet the chance of life forms in each of the many universes is similar to that in our universe (Sec. 15.2). The following chapter is to study the evolutions for such developments.

Part II. ORIGINS

7. THE BEGINNINGS OF OUR UNIVERSE

7.1 The Mechanisms for Survival through Evolution

We have completed our journey out to the farthest reaches of space, so we are now preparing to make the other journey I promised you, the one through time, from the beginning to the present. Before we can start that history, we need two preparations. First, since evolution comes up so often, we shall study what it is, with 12 observations of it, in this section. Secondly, in Sec. 7.2, we will get a feel for the extremely high temperatures and pressures that occurred during the very earliest stages of our universe, according to nearly all models for those stages.

We began this book with the view of evolution as nature unfurling. Darwin had a detailed understanding of that, and called it *descent with modification.* Eric Chaisson described its physics and he calls it **Cosmic Evolution**[1]. He recognized energy flow everywhere as causing continuing increase of complexity in nature, from the beginning of our universe to advanced life forms.

In the present book we found and will continue to note that evolution occurs commonly, like in a universal sense of being everywhere and at all times, with 'universal' being a good word to indicate also that all aspects of the universe are involved, in its inorganic and organic domains. I therefore call that treatment **Universal Evolution.** It differs from that of Chaisson in not resting so exclusively on observations of energy but on all aspects of evolution. We see it from the earliest stages of this universe expanding out, and the expansion forces organization into new forms, continuing into the future; we immerse ourselves in this view and interpretation. Evolution was present all the time (Chaisson agrees with that), because there was nothing to stop it after the very beginning of our basics in the multiverse. Nay, it is the evolution that was **driving** the developments. Evolution did not begin with Charles Robert Darwin in the 19th Century, as so many scholars seem to tell us. He and Alfred Russel Wallace[2] **discovered** it in the 19th Century, as we shall describe in Sec. 11.2.

Historian David Christian initiated a new branch for the study and teaching of history going back to the beginning of our universe, and he called that **Big History**[3]. Another historian, Fred Spier, advances an explanatory scheme for Big History based on the ways in which flows of energy and matter brought the rise of **complexity**[3].

I received some critique that the concepts 'natural selection,' and 'evolution' are for exclusive usage in the biological domain. My answer is that we find natural selection, and therefore evolution, everywhere. Our favorite example of this is Fig. 5.4, referring to it often, of **atomic nuclei within the stars**; there also is evolution **of the whole stars**, depending on their original masses. After evolution originated in the multiverse, it unfurled on to our universe, its stars, interstellar grains, and finally to the solar system, appearing in ever greater capability and therefore in ever greater complexity. Evolution itself evolves, into ever more complex species, of the atoms, molecules, cells and the various forms of life; the tools became more complex in life and its cells, DNA, chromosomes and variation.

The principal characteristic, that all forms and species seem to have in common, is the **survival** aspect. There may be exceptions, reversals, and also jumps, but the development eventually seems to happen to other forms and species, which are more capable and therefore more complex - or more complex and therefore more capable - there always seems to be the survival aspect and we will see it emerging in the multiverse (Sec. 7.3). Any participating universe that is not of the proper mass and physics vanishes; it does not survive. There is a euphemism in saying that evolution is the survival of the fittest, or one could consider it reciprocity, and so it is for Survival through Evolution as well as Evolution through Survival. This an example of using the Reciprocity Principle.

Professor Chandrasekhar was adamant with his students that they understood and applied the **Reciprocity Principle**, and now I pass this on to you. It is easy to find examples in science, and it is fun to find them in daily life. In fact, he asked us in an exam question to mention such examples. I could think only of, "When I am kind to someone, (s)he will be kind to me", and expected a low grade, but he rated it tops as an effective guideline for human and international relations. We saw an example in Sec. 6.6 where the characteristics of the multiverse were found from those of our universe, but it was the multiverse that brought them to our universe.

Please note that the definition for **species is** not limited to the one that is taught in biology, but the more general one, which is also in our dictionaries, as being "a class of entities having common attributes and being designated by a common name". Another term was considered for the inorganic domain such as "kind" instead of species, "descent" instead of evolution. However, the essence of Nature is a single evolution, from subatomic particles to the most complex life forms, and that understanding is enhanced by using the word 'species' for all. Similarly, we will avoid speaking of inorganic versus organic evolutions, because they simply are increasingly progressing stages.

The above definition implies that each species is a separate step. The steps are distinct, greater than the **variations** we observe within the species; in biology for instance when two finches on two separate Galápagos islands have different beaks because different foods brought first the variation about, but now they cannot crossbreed anymore. They are separate species, which is a good thing for the crossbred would starve on both islands. Then a basic observation is of a **quantization** in nature. Multiverse, universe, stars, and even people and animals, come with a range of size for each species, and that range is limited. For atoms, there is no variation, there is no range, they are identical within each element.

The key to accepting such universal evolution from the multiverse to modern society, is the recognition that life forms use the earlier mechanisms *plus* the later ones that brought variation within the species. Each species came from earlier evolutions, while the refinements of variation came later from the *added* mechanisms of chromosomes, etc. The usage of the above definition avoids misunderstanding, it simplifies matters, and the evolution of life fits in well. The opposition is persistent though, remarks have been made like "galaxies do not produce baby galaxies".

In any case, here is a listing of 12 observations that I have made of evolutions for which we have sufficient data, namely the ones of whole stars, nuclei of atoms within the stars, molecules in the interstellar medium, and of those on Earth. They are in logical order but it is not an order in time or a ranking of relative importance. The 12 are informal examples, which may not be complete - you can make them **yourself**, and you can make more. The attributes for life forms are not included, but some appear at the end of this section and in Chapter 11. There are many as the evolution of life occurred with procedures of mating, mixing of chromosomes, variation within the species, and with the need for food and rest, sickness and death. The 12 observations are stated in **bold** face, followed by some remarks. The easiest way to consider and remember them is by relating them to characteristics in your own environment.

1. **Possibilities for combination are tried continuously, in interactive but random "trial-and-error" searching for what might fit, and taking the next step towards a new species whenever such a fit is found.** This observation lies at the root of the title for this book, "Survival through Evolution", because it seems to be the key to evolution, which is seen for Points 6 – 12, particularly 12, and the ones for life, below.

2. **Nature produces energetically and prolifically.** We know how strong the drive can be to produce and reproduce.

3. **The information content is generally too small to have any goal or capability beyond the next step.** The atomic affinities do provide combinations, such as when two hydrogen atoms meet an oxygen atom to form H_2O on an interstellar grain. That information content is probably for not much more than the next step, the next species. In any case, it is out of the question that the earliest particles, or even atoms, molecules and simple cells could carry information to make people, for instance; there was not sufficient information content in the beginning to predict or control the final outcome. There is, in fact, no final outcome. One cannot consider people, for instance, to be in their final stage of evolution (Sec. 11.5).

4. **Evolution sometimes produces misfits, it may be stagnant, or it may make jumps.** The trials of Point 1 do not always come up with what one might consider the best solution. "Best" is a value judgment and usually slanted in an anthropic way, relating to humans; "progressed" may be a somewhat better description.

Wallace, who also was leading the discoveries of evolution[2] agreed with the existence of such jumps, but Darwin did not, surprisingly, "natura non facit saltum," nature does not make a jump. An example of the opposite is the asteroid impact of 65 million years ago which caused an enhancement of evolution (Sec. 12.4). Darwin did point out how misfits appear. There also are exceptions in which evolutionary trends seem to proceed from complexity to simplicity, and there are species adapted to an environment that has barely changed, like sharks that have shown hardly any evolutionary change during the past 200 million years. One of the simplest critters, the bacteria, have survived and thrived the longest, in numbers that are astronomical in a variety of environments.

5. **Among the species before life, evolution proceeds without variation within the species, but uncertainty and indeterminacy are already seen before.** The emergence of new species is a continuation from earlier evolutions, clearly illustrated in Fig. 5.4. It shows the classical example of fusion evolution, which progresses in the steps shown from whole helium nuclei to whole iron nuclei, to whole species. There is **no**

variation within these species; Nature had not yet arrived at the complexity of cells and chromosomes at this stage. The atoms and molecules do not have variation within the species, as is observed in Sec. 6.1. The 150 molecules identified in space show exactly the same spectra in a variety of interstellar clouds at various distances and directions in the Milky Way; the atoms of these molecules must therefore also be exactly same. The variation within living species may have evolved from the properties of uncertainty and indeterminacy in the atomic world (Sec. 8.2).

6. The outcome depends on the environment; natural selection is seen everywhere. The key point of evolution is natural selection, for which Wallace and Darwin are recognized as the discoverers (Sec. 11.2). Others had studied evolution, including Darwin's grandfather, but the emphasis on natural selection made the difference. The earliest evolutions also show natural selection is clearly seen in the earliest evolutions. Figure 5.4 is again our example, showing that carbon and nitrogen are produced at density 700 gm/cm^3 and temperature 6 x 10^8 K, while production of the more complex silicon and sulfur needs the more powerful environment of 10^7 gm/cm^3 and 3 x 10^9 K.

7. The environment is also modified, interactively. This point seems indicative to what physically happens in natural selection, namely the wave structure of matter and the interaction of space waves, whereby the environment and its species interact and develop together. Chapter 8 is dedicated to explaining the physics.

8. Through the changes of species and environment, the system is sustained. The stellar interior is again an example for confirming the observations of 7 and 8. Another example is the event of a baby arriving in the household of a young couple; their environment is modified, and their system sustained.

9. The result of each stage is more than the sum of the components that went into it; there is an apparent efficiency. The new arrival in a family is an example for this because the couple produced together what they cannot make alone, a child. The South African general and political leader, Jan Smuts, brought this surprising attribute to the fore, and it caused quite a stir among philosophers in Europe following through on the word **holism** in the title of his 1926 book[4]. Business schools use the term **synergy** for the related concept of combined action bringing exceptional results; an orchestra may be a beautiful example. Fundamental evidence seems given in the formation of four hydrogen nuclei producing one helium nucleus. If the mass of the hydrogen nucleus (the proton) is denoted by 1 m_p the resulting mass of the helium nucleus is found to be 3.97

m_p and not 4.00 m_p, so that 0.03 m_p is liberated for other action. One speaks of the **efficiency** of the combination; it may be driving evolutions as a self-organization[5], in which Nature seems to perform spontaneously by itself.

10. Much of evolution is unpredictable, but there seems to be an overall trend towards greater complexity; what *can be* produced, *must be*, and *will be* produced. We have seen unpredictability already in the above Points 4 and 5. For interstellar grains, hydrogen atoms meeting two oxygen atoms must combine and result in water, H_2O; there is no choice and no delay.

11. The greater complexity usually brings greater capability, so that evolution itself evolves and accelerates. We see evolution itself evolving to greater possibilities from the combination of existing capabilities - the successive new species becomes more complex and thereby more capable. I mean with "capable" exactly that, the ability to do new things, and no value judgment is implied such as an improvement or aiming at an ultimate goal. The instructions of each stage and species were for them, and the combination into the next species will bring a coalition of instructions, taking the efficiency of the above Point 5 into account, to increase the complexity and capability. Each next step or species becomes more able to do things, more capable, and we thereby see the processes accelerate. The capabilities have increased from subatomic particles to atoms to molecules to cells and from simple to "advanced" life forms. An astounding example of acceleration is that the simple early evolutions on Earth took billions of years, they speeded up gradually, then faster and faster, and finally came surging into its most complex evolution of people, which took only a few million years.

12. Evolution appears to be the essence and guiding drive of the cosmos. This observation is made often and impressively, it is a summary of everything before, and of the ones still to follow for life forms as well. We find evolution everywhere and at all times, affecting all aspects of the universe, and their origin occurred in the multiverse.

Some 2×10^9 years ago the addition of the tools for life forms came with the random genetic **variation** within the living species. Wallace and Darwin were studying the ensemble of evolutions obtained from both the inorganic and organic domains. They did not know of quantum rules or chromosomes. They would of course have been delighted to know more of the combined effects of physics and biology, and that their theory of evolution rests on the one that worked in the multiverse (Sec. 7.3). We shall use these observations to see the earliest stages of our universe in most of the models for such earliest and extreme environment.

7.2 Extreme Pressures and Temperatures

Older scenarios have our early universe originating in a small volume with densities even higher than we quoted for neutron stars and black holes. For a proper appreciation of the compression, we remember that the universe is presently expanding. We make a **thought experiment**, imagining a reversal of the expansion into a shrinking; we try to imagine the history of our universe by thinking in time.

When the universe's volume was smaller, the galaxies and stars were closer together - originally they were a hot gaseous ensemble, a **plasma**. When the earlier density and compression had been higher yet, so would have been the energetic interaction and mutual collision of the components, and the temperature would have been higher through all that activity.

For conditions and environments as in the sun, the electrically charged components like free protons, electrons, neutrons, and photons is known and referred to as a **plasma**. But the earlier compression and energies were more extreme than that and even the atomic nuclei were broken apart. At first, it is not so different from the plasma in the interiors of massive stars, but soon the temperatures and pressures were becoming much higher because the compression was caused by masses no longer as in stars, but by the mass of the **entire universe,** on the order of 10^{23} stellar masses. Our imagination of the history has reached the domain of unique conditions that happened only once to such a universe and its sub-subatomic components; this is where they originated in evolution in and from that extreme environment, at least in these models. The densities were perhaps 10^{70} gram/cm^3 and temperatures 10^{25} K (these values are merely guesses).

If you had been unprepared for high densities, you would simply shake your head uncomprehendingly at such numbers, but we have been leading up to this topic throughout our book. In Sec. 1.4 we began with zero density at the top of our atmosphere and gradually increased it towards the bottom of our atmosphere; for comparison we also considered our own 0.99 gram/cm^3 as an extremely low one due to spacious atomic structure that we will discuss in detail in Sec. 8.3. Continuing with less space between components, we saw 10^5 and 10^{15} gram/cm^3 for white dwarfs and neutron stars, and we finally avoided giving a firm number for black holes because we do not know but surmise numbers like 10^{50} gram/cm^3 depending on the mass.

Anyway, if ours had been a **pulsating** or **cyclic** universe, then at some time the compression and contraction of super-hot and super-dense plasma would have reversed into an outburst and expansion of space[6]. Until the

1990s, it was thought that ours is a pulsating universe, this has no longer been the case for a decade, until a string-theory model appeared in a book in 2007 (Sec. 7.3). Otherwise, by about 2001 there was little doubt that the Big Bang model was correct[6], and it became accepted as the Standard Model (Sec. 7.3). With the thought experiment of this Section we are better prepared to discuss the hot and dense start of our universe and all its basics and physics in that model. The M(α) model of Sec. 6.6 has a later and milder origin of our universe with radiation and subatomic particles from old universes being re-configured into protons and other particles rather quietly, without a big bang, at pressures and temperatures that were less extreme.

7.3 The Origin of Evolution

First, we should mention **string theories**, which were originated to avoid some classical problems of the Standard Model. They facilitate storage of information by having a type of small particle replacing the point particle of quantum theory, this is called a string; a number of theoretical physicists study these topics, without observational evidence thus far. Brian Greene[7] has overviewed string theories and their history. The strings come in various small sizes and shapes for storing varieties of properties. Several **dimensions**, at least five, are added to the four we are used to (which are up and down, close and far, left and right, and for time). Greene dismisses any difficulty in imagining other dimensions and complex particles. The variety of particle properties is a reflection of the various ways in which a string can **vibrate**: the electron vibrates one-way, quarks in another, etc. Greene regales in the analogy with vibrational patterns of music, how the strings orchestrate the evolution of the world into a cosmic symphony. Murray Gell-Mann had made an earlier overview of string theory; he is the inventor of **quarks** as components of particles such as the protons[8].

Lisa Randall wrote a book about string theory in which she discusses larger string particles called branes (like membranes)[9]. Paul Steinhardt and Neil Turok, who criticize inflation theories, present a string theory for our pulsating universe[9] [they assume exactly critical mass, which may be ruled out by a precisely established factor of 1.94 (Sec. 6.6)]. Peter Woit criticizes string theory[9], as others have done, and that is the state of affairs for understanding the earliest stages of our universe. Inflation and string theories seemed needed at the time to explain the origins of our single universe and its intricate particle physics.

This paragraph begins the description of the origins of evolution, first for the **Standard Model**, and after that for the case of the multiverse. In the standard model, our universe is considered to be the only one, without a multiverse, which seems to be an anthropic concept. Furthermore, the words Big Bang are commonly used, which is useful as a quick name for the model, but this also may be an anthropic concept. The words indicate an explosion, an instantaneous event, and that is so if one uses the **second** for the timing of early ages of our universe as $t = 10^{-43}$ and $t = 10^{-35}$ seconds; all textbooks seem to use the second. Such times are difficult to imagine, and it is improper to use the second. It is logical to do so for the timing of human activities, because our reaction time is not faster than about one-hundredth of a second, 10^{-2} secs. However, here we consider the activities of quantum fluctuations, strings, branes, quarks, and others. They are much faster than we are, doing things in times like 10^{-43} secs over typical distances, for them, of about 10^{-35} meters. We must therefore use their unit, which is the **Planck Time (PT)**, the time scale for the most elementary actions of the quantum theory; in seconds, PT is close to 10^{-43} sec, a second divided by a 1 followed by 43 zeros.

The Planck Time should also be used to indicate the epochs. The starting time for the Standard Model is then the epoch of zero Planck Time, **t = 0 PT**; when measured in seconds it would be $t = 10^{-43}$ sec. The first important interval is the one between the epochs of 0 PT and 10^{11} PT; that is between 10^{-43} and 10^{-32} seconds. There are good reasons for choosing these epochs: the former is the beginning, while rather well-known events in particle physics occur at 10^{-32} sec (Sec. 7.4). From the starting time at 0 PT until age of 10^{-32} seconds, it takes $(10^{-32} - 10^{-43})/10^{-43} \approx \mathbf{10^{11}}$ **PT**, one hundred billion Planck Times. That interval allowed each of the components like quantum fluctuations to have 10^{11} activities, 10^{11} fluctuations. That is a lot of activity in the quantum domain, not like an explosion at all, but more like a **slow burn**, and this is still a lower limit because relativity theory indicates that under such conditions of high temperature and pressure, time may have slowed down[10].

My understanding of the events at that time interval is based on the following four assumptions or premises. The first is the above usage of PT instead of seconds. The second follows from thinking in reverse the presently observed expansion of our universe (as we did in Sec. 7.2); it happened 13.73 billion years ago, with the universe concentrated in a small size at high density, temperature, pressure and energy.

The third premise follows from simple logic, namely that all of **our basics**, including evolution, would have had to be produced during this very beginning; the earliest stage must have had or produced the earliest forms of relativity, quantum mechanics, gravity physics, self-organization, evolution itself, all our fundamental concepts. The latter must include everything we are familiar with, such as action-and-reaction, the Reciprocity Principle, curvature of space, the restricted velocity of light, and the 12 attributes of evolution outlined in Sec. 7.1.

The fourth premise is merely an indication, namely that some such spectacular quantum and relativity environment was needed to produce the eventual products, of properties so spectacular, that we will see in Sec. 8.1 with the greatest physicists asking, "can nature possibly be so absurd?" The atoms, molecules, cells and cerebral cortex have ethereal properties that had to have come from some such extreme temperatures and pressures, and a foundation of the above physics.

The conditions of the early high compression and temperature stage are not very well known, in spite of Sec. 7.2, but we do recognize tremendous pent-up **energy**. We also found that a vast amount of time like 10^{11} Planck Times were available for trial-and-error searching for what might fit to make a next iteration, from quantum fluctuations to sub-atomic particles. This trial-and-error searching would have **to happen to come upon** a combination of attributes and properties that would be internally consistent to survive and store information for a next configuration, leading to the epoch of 10^{11} PT (10^{-32} s), for which we know some of the conditions of our universe (Sec. 7.4).

The volume must have stayed small for this stage, without expansion as yet. The condition of this stage is that **interaction flourished** in a no-expansion-yet situation, in which everything could interact with everything else in such confinement. This became clear when the present universe was observed from spacecraft and Earth-based telescopes to be uniform on its largest scales; we shall see this in Sec. 7.4, but for now accept its conclusion that uniformity was needed of the basic components for our universe, which could homogenize only through intensive interaction. However, this leads to an impossible early condition because to interact, the components must be close enough. All of the 10^{23} solar masses of the universe would have to be so confined in the above volume so small that their components could bounce between all of them at the speed of light. Their density would have to be as high as 10^{103} kg/m^3, which is not at all realistic.

At this point of reasoning, the **inflation theories** of Sec. 6.6 took over in the late 1970s, with a complex homogenizing and processing to healthier conditions, and this occurred at about t = 10^{-32} secs. At that time, a fast increase in the size of the universe occurred, an inflationary expansion, which should be followed of course by the continuing expansion we still observe today. The processing was towards the beginning of particle formation, removing a variety of uncertainties in the understanding of the early stages for our sole universe. The Inflation Probe is a space mission especially designed for such studies of that topic and epoch.

The problem with our universe being the only one, is still the question where the 10^{23} solar masses and then their physics came from. There are theories of quantum fluctuations and zero-point energy or electromagnetic quantum-vacuum energy produced by space (Sec. 14.6), but 10^{23} solar masses seems a lot to expect from those mechanisms. The multiverse model has them coming from old decaying galaxies (Sec. 6.6), and for this model the above third and fourth premises are not needed because its powerful and ethereal physics had evolved already in the multiverse.

The multiverse model offers a continuous rebirth of old universes into new ones, with the Chandrasekhar equation in Sec. 6.6 showing the same mass and physics they all share. That **multiverse** then provides a simple example of survival through evolution. To begin with, for any model or theory of a multiverse, its universes should be survivable; they must have the proper mass. If it does not accrete enough material from the inter-universal medium, it will expand quickly into space. Accretion of too much material is not likely to happen beyond the structural limit of the accreted body, but if it should happen, the resulting universe's gravity will make itself collapse. This concept is well known and expressed in terms of a **critical mass**; it is close to the mass of our universe. That calculation is made for the physics of our universe, and, since our universe is a member of the multiverse, the same physics and mass apply to all other universes; they all depend on $M(\alpha) = (hc/G)^{\alpha}$, that is on quantum, relativity, and gravity physics. Furthermore, they are all constituted from the debris of all previous universes (Sec. 6.6).

We found the characteristics of the multiverse from those of our universe. It happened however in the opposite direction in time. The multiverse delivered our universe and **endowed it with its evolution, energy, mass, and physics**. Each universe is thereby capable of doing all that our universe does here. That may include evolution perhaps going as far as having cells and chromosomes as its tools, and making intelligent beings as one of its species, depending on local conditions such as the right distance of their planet from the central star. Let us now follow that evolution at a point in time long after the earliest beginnings, and therefore independent of their various theories and models.

7.4 The First Minute of our Universe

High-energy particle physicists begin to understand the physics of the universe after the epoch of 10^{11} PT, 10^{-32} seconds. Let us briefly overview what they propose in a standard model for the ever-increasing capabilities and phenomena that evolved subsequently.

The following stages did again have ample time for trial-and-error towards the birth of the universe's longer-lasting subatomic particles. The signature of the fiery environs went into the energies that hold the various particles together. The physics for the conditions indicates that particles formed together symmetrically, with particles and anti-particles having opposite properties. The two would largely annihilate each other; when a particle and its anti-particle would meet, there would be **annihilation**. However, it was apparently not a total annihilation. One of the two types would happen to prevail, which is the one we now call the "particles".

Much later, by about $\mathbf{10^{-12}}$ **sec** (10^{31} PT), various nuclear **forces** and their laws are beginning to be recognized in the physicists' modeling of what must have been fabricated at the then prevailing temperature and pressure. By $\mathbf{10^{-6}}$ **sec** (10^{37} PT), there was further adaptation to the subsiding temperature, pressure and density. Natural selection befitting the expanding environment brought the combination of the sub-atomic particles into larger units such as protons, as well as nearly the same numbers of anti-protons with mutually destructive properties. This was always accompanied by annihilation in which they fought it out, the winner had more than the other, and that was the one we now call the **proton.** A large amount of radiation was also produced, the **photons**.

At about **one minute of age**, processes slowed down enough to allow combination of protons and neutrons into larger **nuclei**. The models show a spell of a few minutes during which the conditions were right for the assembly of the nuclei of helium, plus traces of heavier nuclei. However, the change towards less dense conditions happened fast in that expanding universe, and the combination of the atomic nuclei occurred only for those few minutes such that only a limited number of them could be formed. Here lies a success story of modeling because the computations yield a good estimate of the amount of helium nuclei that we presently observe in the universe. The cosmic abundance of helium thus resulted at about one quarter of the total mass, while three quarters remained hydrogen.

About the following years we know little other than that the universe was a plasma of nuclei and electrons, with photons being scattered around, a hot, dense, turbulent and energetic mass in expansion. All we can do now is to join that plasma 400,000 years later.

7.5 The Universe near Age 400,000

A long time passed until the next marked point. Until that time, as far as we know, not much was happening other than **scattering** of light, a situation we discussed in detail in Sec. 5.2 for the sun where it takes a million years for a photon to be scattered and bounced out to the outer levels. It seems clear why it took a spell as long as 400,000 years, because the universe had to come from densities of some 10^{50} gm/cm^3 to the one we are familiar with near one gram/cm^3, or less, our spacious universe.

When the temperature and pressure had reached the level at which scattering diminished suddenly, the great transition took place. Because of the expansion, the density had dropped enough so that the components knocked about far less violently and frequently. The individual nuclei and electrons no longer had enough collisional energy to stay free for their bouncing and scattering acts. The electrons connected around the nuclei to form hydrogen and helium **atoms**. The scattering stopped because the scatterers were no longer free to do so; they were now bound to each other in a spacious configuration. By the time the expanding universe had reached the temperature of about 3000 K, at age of about 400,000, it was no longer an opaque plasma - it had come to the present phase of matter and radiation.

You may have noticed that I paid detailed attention to this stage in our evolution. The reason is that the radiation from that 3000 K stage can still be observed today. It is all around us, far away to where it has expanded since the time when it dropped off clouds of hydrogen and helium that were to separate out into galaxies and stars. Yes, you can say that we are still inside the Big Bang. Its flash is present like an Oort cloud around us but now of radiation rather than of large comets, and at enormously larger distances. That Big Bang had expanded and thereby moved its components farther apart so they could bounce not so vigorously anymore – the cloud has **cooled** to almost absolute zero by now. We call it **the 3-degree-Kelvin radiation**, 3 K being the approximate temperature reported when it was discovered by chance with a radio telescope on Earth. After that discovery, a spacecraft measured the temperature more precisely at 2.725 K – and it was the same in all directions, at least to that precision – it was the Cosmic Background Explorer, **COBE**. The uniformity in all directions is important for inflation theories (Sec. 6.6), but scientists always want to proceed, and spacecraft that are more powerful were therefore proposed.

Later spacecraft indeed study deviations from 2.725 K, small ones, in the fifth decimal place. Those deviations are what the cosmologists were looking for, because the slight mottling shows fragmentation of the original

Big-Bang plasma into isolated clouds of hydrogen and helium, which formed groupings of galaxies and their stars. Physicist David Wilkinson was a strong supporter of COBE and of its follow-up studies so that the next mission was named in his memory the **Wilkinson Microwave Anisotropy Probe** (WMAP). "Anisotropy" (Gk, isos, equal) is the slight unevenness in various directions and distances, as was found.

WMAP makes improved measurements and interesting conclusions through a combination of techniques. An all-sky map shows the deviations, the peaks are 'acoustic' because the expansion and contraction of these density fluctuations are acoustic waves in the plasma, in which radiation pressure competes against gravitational contraction[11]. The celestial map is also analyzed in small area segments, the deviations are plotted against the size of the areas, and certain sizes show peaks. Theory is then fitted to these peaks, varying seven components of the theory, and this is successful because there are many observations, also by other telescopes, such that seven results are obtained[11]. Polarization, measuring intensities of microwave vibrations in up-down and left-right directions, is also studied[11]. The results regarding the 4, 22, and 76% compositions, distance and age are in Sec. 6.5; the **physical essence** of dark matter and dark energy is still puzzling unless the multiverse model in Sec. 6.6 can help. WMAP also derived that the formation of galaxies and their stars began some 200 million years after the beginning of the universe. It was already known that the galaxy formation peaked at a maximum about 5,000 million years after the beginning, which is 5×10^9 years of age for the universe, to be compared with the present age of 13.73×10^9 years.

There are dozens of WMAP members in various teams, and in teams at other telescopes that are used. Three are in Antarctica namely with the ACBAR telescope (Arcminute Cosmology Bolometer Array Receiver), DASI (Degree Angular Scale Interferometer), and by the BOOMERANG balloon telescope of which the remarkable name refers to the atmospheric circulation being such that after launch the balloon returns above about the same place after a few weeks. The results will be refined further by the INTErnational Gamma-Ray Astrophysics Laboratory (INTEGRAL) in space[2]. We return to the amazing atoms that resulted from that spectacular beginning.

8. THE ATOMS

8.1 The Miracles of the Atoms

The evolution near the age of 400,000 was the first into the configuration of atoms, no longer broken up as nuclei and electrons. That combination and its escaping flood of radiation at this age are considered the end of the Big Bang modeling. This is, therefore, the proper place to discuss the atoms, especially because we have just finished the study of the amazing conditions for the origin of their earliest components. We are going to find amazing properties.

Atoms are remarkable in their complexities and capabilities to perform intricate tasks. They consist of **nuclei** surrounded by wave systems of **electrons**[1]; the configuration of the electron appears like a quantum blob surrounding the nucleus, which is routinely shown for the standing-wave patterns made by electrons in a high-resolution scanning microscope[2]. Alternatively, the electron is seen as a particle on the loose, such as charges propagated in electric wires. In any case, the best known is its **electric** attribute we call "negative charge". The electric attraction of the electron and of the oppositely positively charged nucleus holds the atom forcefully together. The hydrogen atom is the simplest, with only one proton as its nucleus and only one electron shell. Larger, heavier atoms have a number of electron shells, with the same number of protons but also with electrically **neutr**al particles in their nuclei that therefore are called **neutr**ons; a carbon atom, for instance, has 6 electrons, and its nucleus 6 protons and 6 neutrons. The neutrons keep the protons far enough apart so that the nuclei do not break up by the protons' repulsive "positive" electric charges.

Another miracle lies in the small dimensions. The radius of the atom is about 10^{-8} centimeters, **a millionth of a human hair**. Atomic dimensions are about a million times smaller than what our eyes can resolve, but electron microscopes allow us to observe the outer configuration of the atoms as is sketched in Fig. 8.1, below. The nucleus of the proton is as much as 100,000 (10^5) times smaller yet, namely 10^{-13} cm.

If our atoms are so small, the number of them within us must be astonishingly large. A nice even number that is easy to remember is **1.00 x 10^{28}** atoms (10 octillion, 10 000 000 000 000 000 000 000 000 000); this is, however, for a rather heavy person weighing 102.0 kg, 225 pounds. You obtain your own number of atoms precisely by dividing your weight, in the buff, in kilograms by 102.0, or in pounds by 225, and then multiplying the result by 1.00×10^{28}. For example, 112.5 pounds yield 5.0×10^{27} atoms, and 155 pounds yield 6.9×10^{27} atoms.

Every one of these atoms has a fabulous assortment of properties, capabilities, and unbelievable energy. Furthermore, this assortment is not just for people, it is for all visible matter of the universe. There still is the other 72% form of energy that the Wilkinson Probe measured. This one is still not understood, but we do have an **energetic universe**[3].

There are **affinities**, attractive forces between particles that enter them into and remain in chemical combination, with complex rules for their application. Atoms combine with other atoms to make molecules, and we denote these in two types: organic and inorganic. The **organic** molecules differ simply by the presence of **carbon**, of which the electron configuration is such that it is a remarkably versatile element for forming bonds. It is from such possibilities of combining that we obtain the variety of molecules important for life -- the carbohydrates, the nucleic acids, and the fats and proteins. On the other hand, molecules classified as **inorganic** are smaller and simpler in structure than organic ones. Inorganic molecules are however also important for life such as water which makes up the bulk of most living things including our own bodies. Life could not exist without an abundant supply of water.

Millions of molecules make up one **cell.** Cells perform vital functions such as taking in energy and expelling waste, controlling movement, transporting food, gases, and waste products, and performing a wide range of special tasks. They are the basic components of life forms and we shall see much more of them in Sec. 11.3. Let us first continue with the basic discussion of their atoms.

The word "atom" comes from the Greek *atomos*, meaning

"indivisible, incapable of being cut smaller (Gk temmain = cut)." For millennia the belief in "atomism" prevailed, which held that the universe is composed of simple, minute, indivisible "atomos" that are **inert**. It is now clear that the universe and humanity are far too dynamic, which would not be possible if they were composed of the inert atoms of the Greeks. Oceans and trees, our thoughts and feelings, making love or hatred[4], even the unpredictability of our actions, all stem from the amazing properties of the atoms within us.

It seems surprising that not until the late 19th Century did physicists move beyond the ancient Greek concept of atoms in order to lay the experimental foundation of our modern understanding. Suddenly, in the **first third of the twentieth century**, great reputations we made by physicists who developed new theories. There were Niels Bohr, Paul Dirac, Albert Einstein, Werner Heisenberg, Ernst Mach, Max Planck, Erwin Schrödinger, and others, for a total of only about 32, all men but for Madame Curie. They had fascinating times but also frustrating times, because it was so astonishingly new to learn and comprehend the esoteric ethereal aspects of the atomic properties. Heisenberg is often quoted on this remarkable history[4].

> I remember discussions with Bohr which went through many hours till very late at night and ended almost in despair; and when at the end of the discussion I went alone for a walk in the neighbouring park I repeated to myself again and again the question: "Can nature possibly be so absurd as it seemed to us in these atomic experiments?"

They would have had fewer hesitations if they had known more about our universe's origin in absurdly high temperatures and pressures. But even the possibility of a Big Bang was not suggested until 1927 and that theory would not reach full acceptance until the end of the century.

We proceed to another property of the atoms: their **long history** and lifetimes. Ever since the last stages of the Big Bang, massive stars have captured and reconfigured atoms in their interiors. Nature is still doing that today. Our present atoms may have gone through several sojourns in massive stars. Violent explosions hurled them out again into widely open spaces where they combine into interstellar molecules and grains over the course of thousands of millions of years. Our atoms later underwent another set of violent transformations when the solar system took shape. That stage was followed by a period of about 4.5×10^9 years during which the

124   THE ATOMS

inorganic and much later the organic components of our planet evolved. Thus, we trace back the history of the atoms currently residing in our bodies. There are even much longer spells of time as laboratory measurements indicate that the hydrogen nucleus, the proton, may have a **lifetime** of at least 10^{35} years, so much longer than even the age of our universe.

8.2 Our World is Quantized

In the following five Sections, we shall on the one hand marvel at the phenomena we observe and experience, but on the other hand admit how for some of them we still wonder about their interpretation.

Let us begin with the fundamental phenomenon of the electron shells around the atomic nucleus, that they can be affected by light, by photons[5]. First, there is the expected effect that the energy of the radiation goes into the electron waves, and that the atomic system gains a more energetic state. Second, a short time later, the excited electron wave may spontaneously jump back to a less energetic state by making a narrowly defined and precisely measured transition. We speak of the latter as being **quantized** because discrete packages, **quanta,** of energy are emitted in the process.

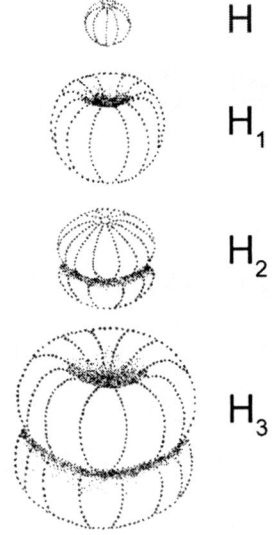

Fig. 8.1. The "ground state" is the smallest, and the other three are successively more excited states of the hydrogen atom.

This is our first encounter with the theory that is **quantum mechanics**[6]. A quantum jump is made with a precisely defined energy, in the form of light at a precisely define wavelength, which shines with a precisely defined color. The light from distant stars thus emits, or is absorbed by its outer layers, in an ensemble of patterns of distinctive wavelengths, and what makes it delightful is that the patterns are recognizably different for the various elements.

Hydrogen, for example, emits a distinct pattern of wavelengths that is characteristic of hydrogen alone, as we noticed in Fig. 5.2 A, and we see here in Fig. 8.1 what the states are between which the transitions occur. Now we learn how these patterns, called **spectra** indicate which elements are present. We have seen in Sec. 5.3 how the overall spectra of the stars can be used to identify different types of stars so that these can be used as standards in the determination of distances for stars.

The quantum jump from the excited electron wave to a less-excited state occurs in about a hundredth-of-a-millionth of a second, 10^{-8} sec. That may seem instantaneous, but only because we use the wrong time scale again. The measure of a second is fine for our daily lives, but too large a unit for an interval in atomic activity; what seems a short time in human experience is a long one in the atomic world. We must use the Planck Time again as we did in Sec. 7.3, and experience that the transition of 10^{-8} sec happens in as many as 10^{35} Planck Times, a long time, for a slow reaction. It is however still difficult to comprehend what is **physically happening** within the atom in those 10^{35} Planck Times.

Some of the puzzles that Heisenberg pondered in the park are still haunting our physicists. They can describe most aspects and phenomena, and be adept at inventing mathematical theories and testing them against observations in expensive laboratories. Engineers are good at using the theories in the design of our modern tools, structures and electronics, compounds and medicines, cars and airplanes that are essential to modern society. But we still have only a partial understanding of what the physical reality is at the atomic level.

The well-known physicist Richard Feynman openly admitted that he did not fully understand details of quantum physics, and he urged his students to let it be[7]. He was referring to aspects such as that laboratory measurements may affect the observation: if your experiment is for observing particles, you find particles, but if it is for finding waves, you find waves. The most puzzling features of quantum mechanics are **indeterminacy, unpredictability**, and an **uncertainty principle**.

For our intuition, it is useful to compare the atomic properties with what we know about our brains, which show **ethereal characteristics** such as the ability to produce thoughts and feelings, memory with integration into intuition, personality, and consciousness, and of uncertainty but also on occasion also a seemingly instantaneous jump of insight into some problem or situation. Without reaching understanding of its details, quantum theory makes sense. We, and the universe, are clearly its offspring - without quantum physics we would not have been here. The properties of our atoms had to be ethereal by high-energy birth and they are within us, for us to gain at least an intuitive familiarity with the quantum world.

We are on firm ground when describing some of the measurable effects of the modern models, and to get a feel for the language spoken in quantum theory as the engineers do when they design our modern conveniences. We have already concluded that energy is of essence to this world. In quantum theory the energy of a photon such as in the above quantized electron jumps, E, is proportional to the **frequency**, f, of the resulting wave; in a relation it is known as $E = hf$, where h is the Planck constant which we will clarify shortly.

Imagine a passing water wave with its series of crests and troughs: the frequency is how often the crests occur in, say, a second, and the distance between crests is the wavelength. It makes sense that the energy is greater when the frequency, that is the activity, is greater; but the wavelength is then shorter because frequency and wavelength are opposites. You can experience that, and the relation between frequency and energy, by watching the standing wave in a rope. Tie the rope to something firm like the wall with one end, and shake the other end up and down. Feel how more energy is required to increase the frequency of the waves, and see how that makes the length of the wave shorter. This experiment can even be done in thought, as a **thought experiment**.

In the atomic world the basic expression is the above $E = hf$, and is called Planck's **quantum principle** because it expresses that when the light is emitted or absorbed, as we discussed five paragraphs before, it is in discrete quanta, whose energy is proportional to their frequency. **Planck's constant** h represents the discreteness. These are simple concepts that lie at the root of quantum physics and therefore at the root of our daily world. Max Planck had a premonition of the latter view in 1899 already when he wrote about his constant h that it[8],

> ...offers the possibility of establishing units for length, mass and time which are independent of specific bodies and which

maintain their meaning for all time and for all civilizations, even those which are extraterrestrial and nonhuman; constants which therefore can be called 'fundamental units of measurement.'

Planck would have been thrilled to know that his fundamental unit of measurement is prominent in $M(\alpha) = (hc/G)^{\alpha}$ for the multiverse, where it may have originated. His perception of extraterrestrial civilizations came into fulfilment over enormous scales, and his nonhuman perception underscores the dismissal of anthropic notions in Sec. 6.6. The most interesting but un-answered question is whether the quantization due to the α exponent in $(hc/G)^{\alpha}$ is connected to the quantization in $E = hf$. Let us continue to marvel at the properties of the atoms, and see the spaciousness of nature again.

8.3 We are Hundred Percent Space

We have noticed how important space and waves are in the universe - they are also important within ourselves. We will discuss the waves in Sec. 8.5, but here we ask, "How much space is there inside me?" The answer is simple if you compare the size of the completed atom with that of the nucleus at its center. Recall that the **radius** of the whole atom (10^{-8} cm) is larger than that of its nucleus (10^{-13} cm) by a factor of 10^5, (100,000). The **volume** goes with the third power of the radius so that the difference in operating domains for the whole atom and for the nucleus becomes a factor of 10^{15}; in other words, there is an enormously larger volume of operating **space** within the whole atom than within the tiny nucleus.

There is a little more space yet inside the atom because the volume of the nucleus is space as well; nuclear physicists are certain of their models for subatomic particles that the proton is not of a material substance, no more than is the electron's standing wave structure. They all are space-wave structures. They represent action rather than solid matter, and we shall discuss this interesting concept further in the next Section. In any case, the fraction of space within the atom is the **whole 100%**. Now comes our staggering conclusion: if the atom consists of 100% space, then so does everything in the universe that is made of atoms.

We can remember that enormity of space for life with another thought experiment. Imagine that the outer perimeter of the hydrogen atom, the electron shell, is as large as the longest outer dimension of an Olympic stadium. How large is then the nucleus of the atom, on center field? As the size of the stadium is roughly three times the length of a football field (300 feet), take the long dimension to be about 1000 feet, which is 10^3 x 12 inches per foot x 2.54 centimeters per inch = 3 x 10^4 cm. Divide that by the above factor of 10^5, and it follows that the nucleus would be 0.3 cm, the size of a pea. All around inside that "stadium" is space. Let us continue on this fabulous track.

8.4 Particles and Laws

The previous Section seems to conclude that we are nothing but space. You will protest that the objects around us are solid and hard, and that we are seen and experienced as people, not space. We are distinct individuals with a great assortment of characteristics, and we experience our mass and weight – how can that be if we would be a void?

However, I did not state that the electron, the nucleus, and the space within the atom are empty, and they are not, not at all. **Space is not a dull void.** Waves are continuously traversing it. You know that from the waves of cell phones, radio and television you can tune onto - such waves are all around us and going through us and the building we are in. There also is the phenomenon that high-energy physicists see inside atomic spaces with their sophisticated equipment. They see **virtual particles** that come and go for short intervals. As a first step towards understanding, you may think that space is like a drop of "clean" water, unless seen under a microscope and then it comes alive. Yes, that is a good start, but those organisms are visitors living in the water, while here we encounter something even more intriguing and more basic: the waves and particles are both part of space; together they are **space waves.**

These are all big words and you may get the feeling that we do not really understand, as Feynman said. You are right, we do not understand yet virtual particles nor really how particles function. However, "empty" space seems to be roiling with quantum activity - particles perpetually coming into existence and then vanishing again, while some like the proton live a very long life. Space has an abundance of ethereal waves and particles. Let us look at particles first. What is a "particle"?

The word "particle" is confusing because it reminds us of a solid grain. But when physicists use the word, they do not mean anything like the Greeks did, an inert grain of dull matter, but rather that **a "particle" is a point in space that produces a phenomenon like mass.** By what it does, a **"particle" is something that provides action and an effect like electricity.** Both boldfaced sentences together may help us understand the particles a little more, as forms of space and waves by which they produce and interact with gravity, electromagnetism, etc. If you have not heard of this before - some teachers do not dare to discuss it - enjoy it as a superb feature of Nature in quantum physics, biology, evolution, and in the interacting components of the universe.

So we shall consider a few particles by what they do, by their **characteristics**. Physicists have long established that particles have a variety of properties such as mass, charge, parity, strangeness, type (baryon, lepton, etc.) and that these properties dictate how they interact. For example, the nucleus of hydrogen, the proton, displays phenomena of mass; so we say that **it has mass**. Mass attracts, as compared to charges that are much stronger, can attract or repel each other depending on which of the opposite charge they have, plus (+) or minus (-).

Einstein noticed that for mass the effects of acceleration are indistinguishable from those of gravity: we experience that when an elevator accelerates and we feel heavy, or decelerates and we feel light[9]. We hear about the astronauts being weightless in steady flight - during take-off the acceleration had brought them a tremendous feeling of weight - and back down on Earth they might stand on a scale again to see their weight. Einstein derived that mass warps space-time and that **"gravity" therefore makes objects follow curved ("warped")** lines when other masses are present. In Fig. 2.1 we saw objects as volumes of mass that warp space such that other objects have special trajectories when they are near enough.

We are indeed one hundred percent space, but we have learned a new meaning of the word **space**, different from the wide-open spaces of Arizona and other common usage. This space is like a medium, active and it contains waves and particles. We have a multitude of particles within us such that we can see ourselves and have a weight on Earth. We do not understand how these particles function or physically inscribe the properties, but we do know that **space-waves** are involved, apparently as concentrations of space and waves, small volumes where waves interact in resonance, and from which specific waves emanate out and to which waves from other particles come in return, so that interactions take place[1].

Such descriptions in science and engineering use rules and **laws**, and we understand that they are based on observations of properties of particles and space waves. **Laws are usually summaries of observations** or of hypotheses to explain observed behavior. They are neither mystical nor mythical rulings of the universe and multiverse. For instance, astronomer Johannes Kepler (1571–1630) published the Three Kepler Laws for Planetary Motion that are listed in our Glossary, but they are his analytical summaries of observations made by his colleague Tycho Brahe who was the expert observer[10]. They are not Kepler's **Edicts** telling the planets how they must behave, like with "Jupiter, Thou shalt move in an elliptical orbit."

8.5 Waves and Matter

If an electron is a wave and/or a particle, then we would do well to consider some of the properties of waves as well. What do we mean when we speak of electron waves and space waves? We are familiar with a wide variety of **waves**: water waves, sound waves, light waves, radio waves, telephone waves, television waves, brain waves, etc. In this modern world, we are surrounded and penetrated by electronic waves day and night. Waves at the atomic level behave somewhat like waves in water, sending ripples of energy from various sources, and reacting in turn to ripples of energy from other sources coming to them. A characteristic of particles is that they indeed have waves rippling forth from them, and the waves are reflected back by whatever else there is in the multiverse; waves from all other particles in the universe are coming in, so that for each particle there are outward and inward waves[1].

Whereas Sec. 8.2 discusses how the energy, E, of an electron wave is related to its frequency f in $E = hf$, we combine this now with the well-known Einstein Equation relating energy to the mass m of the radiating particle, $E = mc^2$ (where c is the velocity of light), so that we can derive an important result. The two relations are valid at the same time for atomic interactions in quantum mechanics, so that we may write $hf = mc^2$. This equality expresses equivalence, meaning that **mass and energy are intertwined**[9]. It is their waves in space that provide all properties and phenomena; this is expressed when we use the combination of the two words in **space wave**. The multiverse, our world, it is all interwoven, interacting, and that brings about the reality of everything in our environment and in our lives.

This conclusion is so fascinating that it is worth saying it again from a different perspective. We encountered spaciousness within ourselves, and concluded that it is a characteristic of all of Nature, of the universe and multiverse. No atoms or nuclei, or any atomic components consist of solidly inert substance; there are no grains, there is nothing of substance within the atom. However, the spaces in the atomic world are not empty voids, they are frequented by fields of waves and particles. We are going to see it once more, in a third way.

8.6. The Cosmic Symphony

Physicist Ernst Mach[11] was aware of the interrelation of matter and space already in 1883, and we now refer to his awareness as **Mach's Principle** saying that space, which is the arena in which matter interacts, is itself an aspect of that matter (the Glossary has more). Interwoven with these concepts is that the properties of atoms and people and of all parts of the universe are **known through their interaction** with the others. We see this again in human relations, because we know each other - and even ourselves - through interactions with the others and the environment. They have also the property in common of dependence on the square of the distance. We experience that property in our daily lives, how the closest encounters are the strongest.

On the one hand, our atoms have the internal spacing they obtained from the universe when it was at the age of about 400,000 years; on the other hand, that epoch itself followed from the properties of protons, electrons, and neutrons made earlier. It is all **interrelated** through interaction. We, the people, are connected with the universe and with other people. This is not a vague platitude but a firm reality executed by the wave properties of our particles and the many more in our local and global environments. These restless atoms are within us and around us, making up everything in the universe. Everything you discuss or see, and the light by which you see it, is a manifestation of the quantum properties that govern reality at the atomic and subatomic levels: the room and everything in it, machinery, and the plane overhead, the atmosphere and its pollution, and people's bodies, thoughts and awareness.

Descriptions in atomic physics are in terms of the acting components' relationships in space and time closely interwoven with their environment; the term **space time** is also used. Not only can electrons change their energies from one wave pattern to another - the overlapping of such waves from all electrons nearby can produce interactions and connections among the atoms. The world is a **flow** of interconnections, for atoms to manufacture molecules and ever-larger associations for molecular strands and cells and all the forms and objects familiar to us. This insight helped to resolve some of Heisenberg's initial frustration with the weirdness of the atoms:

> The world thus appears as a complicated tissue of events, in which connections of different kinds alternate or overlap or continue and thereby determine the texture of the whole.

We have seen string physicist Brian Greene celebrating the analogy with vibrational patterns of music, how his string particles orchestrate the evolution of the world into a cosmic symphony. There may have been powerful sound waves also in the early universe[12]. Our definition of evolution, having the universe as a sequence of interrelated phenomena, underwrites these concepts. We may celebrate it by closing our eyes and imagining within and around us the atoms whirling and twirling in vibrational patterns, orchestrating the cosmic symphony.

9. STARS AND NEBULAE

9.1 Galaxy Formation

Our "Milky Way galaxy" probably formed from a rather spherical configuration into one with a **flat spiral disk** that we observe in Sec. 6.2. So now, we need to answer the question: why are these galaxy disks flat? Is there a connection with the solar system also mostly lying in a flat disk? Figure 9.1 shows in four sketches what happened.

1) We begin with clouds of appropriate mass and size of primarily hydrogen and helium gas that emerged well after the 400,000-years stage of the origin of the universe. It seems logical that the force of gravity in any concentration of matter points towards the center of the cloud, which is how we understand that it is **spherical**. A somewhat higher density is already building near the center.

2) Every cloud in space will rotate; the chance seems zero for having one with zero rotation in this dynamic universe, in which everything is moving and colliding. It is important to appreciate that the rotation has a centrifugal effect, a flight-from-the-center like you feel for a weight tied to the end of a rope when you swing it around. This will tend to make matter fly away from the axis of rotation seen in the figure. Near the poles of the rotation, however, there is only gravity pulling towards the center, and no flight from the center. The effects on the whole nebula, the combination of the gravity and flight from the center, tend towards a flat shape as is seen in the Figures.

3) In the third figure we see the result, with the mass further gravitated in a thin plane, the **galactic plane**. There may be some residual gas colliding with local concentrations, especially near the galactic plane where there is more material, and the gas will then nearly stop in its track and thereby further enhance the mass concentration towards the galactic plane

4) The bottom figure is for the solar system, with its **ecliptic plane**, that we will discuss in the next Chapter. The importance here is to show that for the solar system the first three stages also apply, starting from the solar nebula, which probably also was rather spherical to begin with.

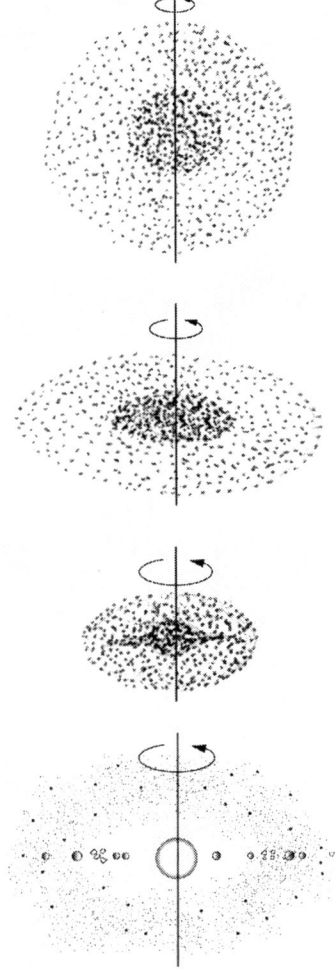

Fig. 9.1. A rotating interstellar dust cloud contracts gravitationally in the vertical direction, the direction of the rotation axis. That is not the case so much in the horizontal direction, the direction of the central plane, because of centrifugal forces. The bottom sketch is for the solar nebula and the resulting solar system.

Formation of stars begins as soon as the local density anywhere in molecular clouds is high enough, as we shall see in detail in the next Section. It is noted, however, that some star formation may have happened already before galaxy formation. Anyway, perpendicular to the galactic plane motion of matter is especially likely to cause star formation from the enhancement of material in the galactic plane. The newly made stars are then likely to keep some of that motion perpendicular to the galactic plane, since they are much smaller than the clouds, and they may then not be stopped in their tracks. This applies also to globular star clusters, and they do show motion perpendicular to the galactic plane. Their formation must have occurred early, making stars that are now old, but the process kept up into more recent times making stars that are not so old. Globular clusters therefore have a **variety of ages** as has indeed been found, in a confirmation of these concepts[1].

Nebular and galactic collisions play a role in the evolution of galaxies[2]. Collisions of whole galaxies must have happened, and they are still occurring. It is merely an example as **collisions are to be expected everywhere** in a universe with restless motion. The three types of galaxies (irregular, spiral, and elliptical), show an order of evolution (Sec. 6.3). This is due partly to collisions and partly to the irrevocable aging of stars by the transition from hydrogen and helium to heavier atomic nuclei. We now believe that the irregular galaxies are the original ones. They evolved into the spiral galaxies, with star formation especially occurring within the molecular clouds of the spiral arms. A variety of observations and computer models indicate that the **peak in star formation** was reached near $t = 5 \times 10^9$ years' age of our universe. The elliptical galaxies are finally the graveyard, as the white dwarfs are for solar-type stars. We shall look more in detail at the evolution of stars, which may actually have started forming before the galaxies.

9.2 Star Formation

The **spiral arms** now are the main regions of star formation[3]. Recall that the young massive O and B stars occur at the blue end of the HR diagram, so you would expect that the spiral arms are rather blue, and images of spiral galaxies in visible light show that color.

Molecular clouds are the birthplaces, but their density increases are needed because the clouds as we see them in our skies are generally too tenuous for making a star. The molecules must come closer together before they are mutually attracted by their weak gravitation; a strong increase in density is needed before the cloud can contract by its own gravitation. In the galactic plane there are **density waves,** which are the cause of spiral structure to begin with, but they are sustained by the star formation they cause. The process is a little like the density waves when a freeway becomes too crowded. The traffic may be going at full speed for a while and then suddenly there is a stoppage... it speeds up again... slows down again, etc. - police call it the **caterpillar effect,** that happens only when there is too much traffic. In the galaxy, when the density happens to be high, there will be a piling up of dust and gas and the situation becomes collisional and aggregating[3].

In addition to the density waves, there may be compiling blasts from an exploding O-type or B-type star. The supernova explosion provides **two shockwaves,** first from its flash of radiation coming out at the speed of light, and then the slower and heavier burst of matter of nuclei and electrons such as in a solar storm but enormously more forceful. Now the gas consists no longer of only hydrogen and helium molecules, but also of the heavier molecules such as carbon. Interstellar grains are of course also present, the result of an aging process we shall see more in detail.

Even though the clouds of interstellar gas and grains that we observe in our skies have low density, they have enormous sizes and volumes. Their total mass is large enough for the formation of a group or a **cluster of stars**. Young stellar objects that are not even 100,000 years old can already be observed with X-ray spacecraft so we can study their earliest stages of aggregation and development. Rotation, turbulence, fragmentation, gravitational collapse, and even magnetic fields are all part of stellar evolution.

Let us enjoy a scenario for the formation of an **open cluster** as we learned from all these observations and from detailed computer modeling by several investigators over the past decades. An interstellar cloud of some 3,000 solar masses starts it off at a diameter of 40 lightyears and a low temperature, not much higher than is generally the case for objects in interstellar space, near 10 - 50 K. The first contraction, caused by a density wave and/or by supernova blast, makes the molecules collide more than before so that the temperature rises. A **cooling stage** is then needed for the shrinking and accumulating to continue, and such loss of heat is indeed possible by **resonance emission** from molecules such as carbon monoxide, which is known to be present from observations with astronomical telescopes. To understand the mechanism, please recall what we discussed for cell phones, and now we see the opposite, what the provider does with his transmitter is resonance emission.

All processes happen slowly because of the low temperatures, and it takes some 10 million years for the temperature of the cloud to have dropped to 10 K again. In the process, the cloud's diameter has shrunk to 10 lightyears. The procedure of shrinking, heating, and cooling will repeat, and so this large mass continues to become smaller and will eventually start contracting by their own gravity to greater density, at which fragmentation occurs into separate proto-stellar clouds. The newly formed objects begin as **protostars**[3], a logical name. They are objects that are starting to yield pressures and temperatures high enough to forming O and B stars and their nuclear energy generation. It takes some 10^7 years in a star like the sun for its temperature to go up to several million degrees, at which nuclear fusion sets in. Within O and B stars, however, everything happens faster because of their large masses; it takes them only a few ten thousands of years. After the stellar temperature has become high enough, nuclear energy generation will continue for millions of years in the O and B stars, and billions of years in later types as the sun.

Let us explore more in detail what happens in the interior of a star when its material assembles and its formation begins with the onset of fusion. **Computer modeling** of stellar evolution is a well-established discipline in astrophysics; investigators know gravitation and nuclear physics to compute the pressures and temperatures precisely at increasingly deeper levels towards the interior.

As we look deeper into the star, the pressure increases, because all the material above is gravitationally pressing down towards the center of the body. The atomic particles bounce furiously against each other and that brings even higher pressures and temperatures. Near the center of the star, in the core, the pressure of all this material is so enormous that the temperature rises to millions of degrees. At such pressures and temperatures, even the nuclei of the atoms break up into sub-atomic particles by the energetic mutual collisions. The **evolution of new nuclei** begins. The protons and neutrons are re-assembled from the sub-atomic particles into new and more complex groupings. To begin in outer layers of the star already, four hydrogen nuclei, protons, may combine to become a nucleus of helium; this is **fusion**. Deeper inwards, the higher temperatures and pressure produce larger nuclei as is shown in Fig. 5.4. But not all the mass of the component protons and neutrons is needed, as we mentioned in Sec. 7.1 in terms of a remarkable efficiency, and the excess mass is equivalent to energy in the form of radiation. The pressures due to radiation, **radiation pressure**, and the compression of the gas match the gravitation pressure; the match sustains the system and the star stays alive, shining.

Eventually, however, the fuel supply of a massive and therefore fast-burning star runs out, in the core first, the radiation and its pressure diminish, so that they no longer support the layers in their place. Because of reduced energy and heat generation, the core first **shrinks chaotically** by gravitation, but some other outer fuel will fall inward, and the temperature may temporarily flare up enough for the fusion to heavier elements to occur at the center. The burning will shift away from the very center to a shell farther out. The central parts of the massive star thus burn out first. In rapid order the following occurs: instability, implosion, and violent explosion for supernova events to shatter and disperse their atomic nuclei and electrons into interstellar space[3].

9.3 The Origin of our Atoms

The explosions that shed material out into space are part of the great evolution and life cycle of the universe. The cycle closes upon itself with the formation of new interstellar clouds, which in turn will form new stars to start over again. Notice then an **aging effect** through change in composition from the original hydrogen made at the beginning of the universe to the later and latest products of nuclear fusion processes, that is from hydrogen to helium to heavier nuclei up to iron (Fe) as Fig. 5.4 shows. Note also that the stars and the galaxy as a whole thereby evolve and age together, from being young with hydrogen and helium to gradually getting older with heavier elements.

All of the heavier atoms in the universe - and those within people, too - have experienced at least one and more likely several, of the above cycles of star formation and supernova explosion. After several cycles, the enriched material may have gone into the solar system. Our existence on Earth depends on this because we need and **consist of the heavier elements** like carbon, oxygen, and nitrogen in addition to the hydrogen for water, H_2O. Our complex forms of life require heavier elements; life could not have formed from the primordial hydrogen and helium. We have seen that the space density of the early universe decreased so rapidly that there was hardly any time for heavier matter to form then, but all our hydrogen and helium did emerge with a few traces of heavier elements as well. Much later, most of the heavy nuclei emerge from the O and B stars. These massive stars distribute their material into space because of their short life and supernova explosion, and the tenuous interstellar space thus becomes gradually seeded with all the elements[3].

Thus we see again, as in Sec. 7.1, interaction with the **environment** as well as natural selection because of what is formed and can survive, depending on density and temperature. For atomic nuclei, the early universe had produced the supply of all the protons, neutrons, and electrons. The stellar stages, that were much slower, toiled to yield the array of nuclei of the chemical elements up to iron (Fe in Fig. 5.5). For the origin and natural selection of some of the very heaviest elements, heavier than iron, special conditions occur during supernova explosions or gamma ray bursts. These large nuclei cannot produce yet heavier ones and they have a less-stable and larger configuration than the ones of Fig. 5.5.

We also see the interplay of action with its duration. Most violent and short-timed was the formation of atomic constituents during the birth of the universe, for which its action intervals are measured in **Planck Times**. Atomic fusion within the stars is much less violent; the times for this stage are measured in **millions of years**. In the next section, we shall see equally important actions for the formation of our molecules. Molecular bonds are much less forceful than the atomic ones, and we will therefore see that gentle but cold environments suffice, taking truly long time spans of **billions of years** in the interstellar medium.

9.4 The Origin of our Molecules

The next stage in our evolution and in our trip through time is the intricate fabrication in the interstellar medium (ISM), where we have seen in Sec. 6.1 how **molecules** are being discovered and identified in gas and grains. The dense clouds of hydrogen and helium atoms from the early universe produced the original ISM; this was interstellar **gas without grains** at first. However, the O and B stars were also forming, at first only from the hydrogen and helium nuclei, which came equipped with their universe's in-built characteristics called "quantum numbers" and "selection rules". Now there must not be any original hydrogen-and-helium O and B stars left. These produce the heavier nuclei; Fig. 5.4 displays the natural selection and stepwise evolution. That brought and now sustains the evolution of the ISM. The O and B stars started to fling their elements out in supernova display and violence, way out into space. Within millions of years, the interstellar clouds were being enriched with atoms more complicated than those of the original hydrogen and helium.

It is therefore in the cold interstellar space where we find that next larger species in nature - the molecule; there are in space large but tenuous gatherings of molecules in **molecular clouds**. When two hydrogen atoms should meet, wherever they meet, they must immediately form the hydrogen molecule. Atoms of all other elements meet and join, they make all possible combinations of molecules; Table 6.1 shows some of them and their discoveries in space.

Another miracle is that the accumulation of molecules can make **grains** in the molecular clouds, an astonishing process, and phenomenon that no one had expected, no one had predicted in advance. Even its discovery took millennia. For thousands of years our ancestors had been looking at these clouds, as you did in your naked-eye observations of Sec. 6.1, and seen so beautifully in Fig. 6.7. However, the desert dwellers and later the astronomers did not understand their observations as a partial blocking of starlight. Gases could not do that, and no one thought of solids in space. This haziness went on until the 1930s, when astronomers had to conclude that their determinations of stellar distance were sometimes terribly wrong. I am referring to the comparison of apparent and intrinsic brightness as we discussed with the HR diagram in Sec. 5.3. Star brightness, luminosity, was being lost, stars were too faint and therefore interpreted to be farther away than they actually were when the distance was surmised by other conclusions. Astronomers lost their reputation over this problem. Finally, the conclusion emerged that these clouds had to have grains of a size that **resonates** with the light waves, as we explained in comparison with cell-phone reception, and is therefore **effective in the scattering and absorption** of light. To do all of that, their size range must be near the wavelength of visible light. This would make sense, again comparing with the cell phone. Thus it was finally understood that there is an unbelievably large number of grains in space with sizes about 6×10^{-5} cm, 600 nanometers (nm). These parameters and some compositional ones were then determined with instruments for measuring brightness and polarization[4] of the scattered light.

Interstellar grains is the name for such accumulations of molecules; "interstellar particles" is also used, but I will not do that as there is already enough confusion about the word "particle". The molecules and grains both come about thanks to an aggressive tendency for atoms and molecules, to bond together into whatever combinations their electron structures allow. It is not merely that they *can* combine, they *have to* combine; the molecular affinity and bonding laws are known in chemistry. For our tour through time, the information content in universal evolution has come to the molecular level. It is also amazing to realize that the atoms that land on the surface of the grain must be surviving the impact and then **migrating** over the surface in order to be able to meet the others; they may be pushed around by the impacts and by stellar radiation and winds (Sec. 1.2 has the example of the solar wind).

Interstellar grains may also have an intriguing starting mechanism because **condensation nuclei** appear needed to get a grain to grow. You can observe an example again with the naked eye, but now on a high airplane in the daytime. The high terrestrial atmosphere may be saturated with water vapor while still no droplets will form until there are hygroscopic, H_2O-attracting, substances that will function as a nucleus for the droplets and get the process started. When a plane flies through, the soot from its engines will do the job, and the condensation trail, "**contrail**", forms immediately. From a window inside an airplane, behind the jet engine, we can sometimes see the contrail being formed. We can often see, down below, the shadow of the contrail drawn by our own plane (provided we sit on the side where the sun is not, of course). If the humidity is high enough, the clouds will continue to grow; they can even make it a cloudy day when many planes are flying, like in Western Europe where its cloudiness is made worse by air traffic. For the condensation nuclei of the interstellar grains, the extended atmospheres of large and cool stars expel small carbon and graphite particles. These are the giant stars of type M on the right in the HR diagram. Graphite and carbon are "hygroscopic", *i.e.*, they stimulate growth by their attraction and absorption of the interstellar molecules.

Once started, the settling process continues so that the cores have mantles of various ices such as those of water, methane, and ammonia. Some heavier substances, even some metallic atoms, have probably frozen into the mix as well. It is **cold** in the interstellar clouds, not much above absolute zero, so the pace of action is **slow**, but there is plenty of **time** available. The ISM processes have billions of years to go, however slowly, for each grain to capture as many as ten million molecules for its formation.

This evolution shows natural selection, and we recognize other **evolution attributes** of Sec. 7.1. When the interstellar grain drifts too close to a star, the stellar energetic short-wave ultraviolet radiation will break up the atomic and molecular bonds, and the loose molecules on the surface leave the grain. Or, the opposite may happen if the radiation is less strong, a stimulation of grain growth, as the loose molecules migrate over its surface, pushed around by the radiation. Collisions among the interstellar grains also occur with the result that they may be broken up; but after each one of these events, the process starts all over, and new combinations may come to the fore, in "trial-and-error search". This may bring homogenizing, making the interstellar medium more uniform, as we saw for the inter-universal medium. Interaction with the environment is repeated over-and-over.

Atoms form molecules on or inside the grain, while through that process the grain grows. The environment makes the molecules, and the molecules make the environment. The system sustains itself. The enriched, combined and possibly already somewhat stringed assortment of molecules happens because of the long times for evolution, stellar radiation is available for their activation, and all the types of atoms and already produced molecules will come visiting. Nature produces prolifically. Adaptive evolution always finds a combination to go on, to survive, randomly at work, without a pre-set long-range goal. Thus was the material fabricated for humanity's abode, the solar system.

10. OUR SOLAR SYSTEM

10.1 The Solar Nebula

We continue with another case of formation from a nebula of the intergalactic medium as we did for our galaxy in Sec. 9.1, namely a much smaller cloud of the interstellar medium from which our solar system formed. Another major difference is that the interstellar medium had atoms larger, well beyond even Fe, than those of the original hydrogen and helium. The transformation inside O and B stars had been going on for some 9.2×10^9 years as the formation of our solar system started **4.5×10^9 years ago** (taking the age of the universe of 13.73×10^9 y into account). It began with an interstellar cloud that we call the solar nebula; the philosopher Immanuel Kant (1724-1804) had already proposed an origin from such a nebula, and several scientists have improved that model[1].

Our tenuous interstellar nebula must have received an increase in density towards a collapse by self-gravitation, and there is evidence that the push came from a nearby supernova. The Allende meteorite of Sec. 2.1 particularly shows its effects[2]. Its shock wave must have caused the contraction of our nebula by pushing material closer together, and it eventually became dense enough for further contraction by **self-gravitation**. The estimated radius of the nebula had at first become about a quarter of a lightyear, 17,000 AU, and the mass a few times that of the sun. Mass loss occurred due to collisions and evaporation, and eventually the total mass would be close to that of the sun of course, and the radius of the nebula would have shrunk to the 50 AU we mentioned for the boundary of Trans Neptunian Objects in Sec. 4.6.

It must have been a chaotic world of gravitational surges and whirling fragments towards a variety of separate volumes for celestial body formation, for the sun first and for the planets later. The greatest concentration occurred near the center of the nebula, where a turbulent interplay of gravitation, radiation, and magnetic fields formed the early protostar that would evolve into being the sun. This happened rather quickly - astronomically speaking - within only a few million years of **pressing and grinding turmoil** within the dense cloud into a regular configuration of a star with eventually its planets. The nebula made the changes, and changed in the process – we continue.

10.2 Early Years

There are five successive evolution stages for our solar system; for preparation, please look at the overview in the lowest part of Fig. 9.1. Stage 1 is the one of the previous section, beginning at the place near the galactic plane where enough gas and dust happened to be present for formation of the system, as the Figure shows. We know from the present configuration that it must have been a **rotating cloud**, so that we see in Stage 2 the centrifugal effect that tended for the material to fly away from the rotation axis, but gravity tended to draw the material towards the center, and Fig. 9.1 shows what happened in its first three sketches.

All effects changed with time. When gravity drew some material closer towards the center, the rotation speeded up; this is a well-known effect for ice skaters when they draw their arms in closer to the body. The result of all this was a mass at the center so concentrated that atomic components were broken apart, fusion set in, and it became an early version of the radiating sun, still settling in. Farther outward, an extended discus evolved in the plane of the solar system, **the ecliptic plane**, with separate clouds and eventually their objects orbiting about the central body.

Stage 3 happened next in the ecliptic with the material repeating the accumulation procedures many times over into blobs that we call **planetesimals**, "little planets". Near the sun, the planetesimals were hot (remember the high surface temperature of the planet Mercury). Any ice, snow, and gas substances of the original nebula blasted or evaporated away by solar storms and radiation. Only the sands, stones, and silicates remained, as we know from our deserts and the barrenness of Mercury, Moon, and Mars, while we have relatively few of the lighter gas, snow, and ices left. Farther away from the sun, the temperatures were low, such that the gasses, ices, and snows of the giant planets, satellites, and rings could survive, as well as the comets. The whole system generally cooled by radiating its heat away. Then we see in Stage 4 the further development of the planetesimals colliding and growing together into relatively small planets close to the sun. The larger structures of the giant planets Jupiter, Saturn, Uranus, Neptune and their satellites developed also, as we shall see next.

10.3 Later Times

Stage 5 is the final evolution, when completed planets and asteroids were left in the ecliptic mid-plane of the newly formed disc, while the smaller objects were dispersed by collisions into all directions, and they mostly cleared away from the system. As for the budding sun, whenever its mass became more concentrated, its rate of rotation increased by the ice-skater effect again, and this process may have caused whirls within the material and eventually a break-off of the planetary nebula from the sun. At first, some material surged in gravitationally towards the center, but now some of it also rushed outwards by the increasing radiation pressure of the forming sun. Seen from the outside, our solar system would have appeared as a **violent whirlpool**, with its material collapsing, colliding, and some of it flaring out into space. In the outer parts of the system, the clearing may have occurred still later, as we shall learn soon from the astronauts.

That central body of the solar nebula became an early form of star that was not immediately yet alike our present sun. The temperature in the collapsing cloud and of the young star kept increasing at first, as infrared observations show for similar stellar systems during their own formation. The variable star T in the constellation Taurus (the bull), appears to be in such an early stage, destined to become a star like the sun; in the early sun's history, this is its **T-Tauri stage**. Eventually the sun flowed into its longest stage of life, its present form onto the main sequence of the HR diagram. It had arrived there from the right and below in the diagram, coming to its present brightness and surface temperature of about 6000 K.

While the main part of the solar nebula at the center made the sun, smaller concentrations formed the planets, satellites, asteroids, and comets all the way out to some trans-Neptunian objects. The concentrations of the nebula that were to become the planets were spaced in a regular progression because each had its own **ring of influence** from which its gravity attracted and accreted its material. [The IAU uses the clearing and emptying of the orbital zone as the criterion for a planet (Sec. 2.2)]. The giant planets Uranus and Neptune had begun to form too, but their completion may have taken some hundreds of millions of years in the colder slower outer nebula.

The astronauts brought rock samples from the moon and these caused a big surprise, namely that between 3.95 and 3.80 x 10^9 years ago there had been a **second heavy bombardment,** of meteorites and asteroids. This may indicate that the formation of Uranus and Neptune occurred later, when most of the solar nebula was already clearing[3]; perhaps this supports the new findings that Uranus and Neptune are appreciably different from Jupiter and Saturn. A lot of the material left over from the solar nebula would have been perturbed by the two newly massive outer planets with the result that cometary objects flew off in all directions, towards the inner solar system as well as towards the Oort Cloud of comets, as we discussed in Ch. 4.

The primary satellites of Jupiter and Saturn formed within the local mass concentrations of their planets, in a similar way as the planets themselves grew about the sun. This **nebular model** of solar system origin rests on the information about how the planets and satellites are orbiting in the same direction about sun or planet, and rotating about their axes with that same sense of rotation. There are a few exceptions as we have seen for outer satellites such as Triton orbiting around their planets in the direction opposite to that of the regular satellites. The exceptions may be individual planetesimals that had formed elsewhere, and became locked in later with the gravitation of their planets; we discussed the capture of Triton by Neptune in Sec. 4.4.

Comets and asteroids eventually cleared out of the large open spaces between the planets, but not everywhere. Still left behind are appreciable numbers, mostly stored in their own domains, from which a few occasionally escape through collisions and gravitational perturbations especially by Jupiter. The comets, **"dirty snowballs"**, are in the outer parts where it is cold, except for the ones we see occasionally because they have been perturbed inward and are active with coma and tails. Some of the Centaurs and trans-Neptunian objects are intermediate comets in temporary storage, in transition from the Oort Cloud towards the inner parts of the solar system. Jupiter has locked objects from beyond Neptune into a **Jovian family of comets**, not far from the ecliptic plane. And then there is the asteroid belt, mostly dry and dusty critters because they are closer to the heat of the sun, but their population is not entirely stable and we will see how some may even collide with the earth (Sec. 12.4).

There was a lot of migration and collision, and there still is some of that today. A case of **major impact** may have been that of Uranus, yielding its striking characteristic of a rotation axis that is more than 90 degrees off from the usual one of the planets. The collision may have happened at any time in an impact of a large Centaur or TNO in its "chaotic" orbit caused by close encounters with the major planets. Or, Uranus may have suffered an impact by a giant object at an early stage when the population of objects was still dense, similar to the case of the Earth-Moon system that we shall see next.

10.4 The Origin of Earth and Moon

Earth and Moon share a special history invented only a decade or two ago[4,5]. Before that, scientists simply had no idea to explain a phenomenon that was well established and widely known, that the overall density of the moon is 3.3 grams per cubic centimeter - much lower compared to that of the terrestrial planets, such as 5.5 gram/cm^3 for the earth. This was the **density problem** of the moon. The new and spectacular explanation is that the earth suffered a glancing blow from an object perhaps as large as Mars, which scooped and blasted off from Earth's upper layers enough material to gravitate together again, not far from the earth, in order to form the moon, while the earth gravitated back into spherical shape again.

One might think that the **probability** of such chancy glancing aiming would be too small, as it would indeed be now, but the event would have happened during the early days of solar-system formation, when the solar nebula was still crowded with small and large objects. Planet-building impacts and catastrophic events must have been common during such early stages, as we can see for the lava basins of lunar maria (Sec. 2.3). As for major encounters, they have been mentioned in three other cases, namely for Venus (Sec. 1.4), Uranus (Sec. 4.3) and Pluto (Sec. 4.6).

This particular Earth-Moon model is based on the composition of the moon being similar to that of the earth's mantle, which does not have the differentiated **heavier substances** anymore, because they had sunk down into the early and partially molten terrestrial body (Sec. 1.2). The scooping from the lighter top layers of the earth, having 3.3 gram/cm^3 today, explains the overall lunar density of 3.3 gram/cm^3. The lunar density is the same as that of the earth's surface, in contrast to the high densities encountered in the inner regions of the earth. Let us see this in more detail.

Before the impact that made the moon, the earth had grown by the accumulation of gas, dust, and planetesimals[5]. The objects had fallen in because of the earth's gravitation, and the speed of these falls continually increased because Earth grew larger and therefore gravitationally more powerful. The clobbering together through mighty impacts increased the temperature, as well as radioactive heating did, so that partial **melting** occurred. Then there was the **differentiation**: heavy materials like iron settled towards the core, while the lightest sandy material floated on top. This outer floating crust eventually cooled and hardened, while parts of the interior remained liquid as they still are today, shielded by the mantle from evaporation and rapid loss of heat. Tectonic **plates** solidified, while some liquid or plastic material remained below, such that parts of the surface are still drifting on partially molten layers[5]. We are now going to study the life on and around those plates.

11. LIFE ON EARTH

11.1 The Foundations for Life on Earth

Here comes an integrated story of how the early accretions of gas and dust and of comets and asteroids explain not only the origins of Earth and Moon, but also of life on Earth. Substances that were collected from the solar nebula during the earliest stages of the earth have been covered up by later accretions in a protective layering, and now they may be coming up from deeper down through volcanic action and from the vents and underwater plumes in oceanic ridges. That seems to be evidence for a **deep and hot biosphere** in the terrestrial crust[1].

The gases that erupt from **volcanoes** and **hydrothermal vents** also brought up water (they still do), which was going to play a key role in the early developments of life on Earth[1]; in addition, water may have come from the cometary ices[2]. Next to appear in abundance were carbon dioxide, sulfur, and nitrogen. The deeper layers that had been deposited by comets, asteroids and meteorites were probably too hot for molecules as complex as methane (CH_4), and ammonia (NH_3) to survive[1]. The heat may have broken them up, dissociated them, and the carbon was mostly absorbed in the water and in the rocks, while the nitrogen would eventually make the earth's atmosphere, with the oxygen that would be produced in interaction with life forms later.

Until the end of the second heavy bombardment by meteorites and asteroids of Sec. 10.3, the surfaces of Earth, Mars, and Venus must have been at least partly molten by the energy of the impacts. But the supply of comets, asteroids, and meteorites in interplanetary space was now mostly used up, and the surfaces of planets had a chance to **cool and solidify**. As the outer Earth cooled gradually, more of the water could stay on the surface without evaporating, and there already was a preliminary atmosphere such that the good greenhouse effect of Sec. 1.4 kept the temperature up and the water would not freeze, thereby forming oceans. It must have been a muggy muddy sweltering world with lightning and thunder and volcanoes belching gases, dust, rocks, and steam.

Thus we get to the long stage of still inorganic chemical evolution on Earth, lasting at least 3×10^9 years, three **eons** (geologists like to think in terms of eons). But then the organic evolution tentatively began some trial-and-error for possible combinations, assembling cells as well as molecules, occurring with ever fuller display of the capabilities that could be brought forth. Such thoughts are prominent in Lovelock's **Gaia hypothesis** (after the Greek goddess of the earth Gaea), as a self-regulating system in an integrated evolution of all its living organisms and their environments[3]. The possibilities probed continuously, first in the oceans, then on the wet clays of the shores, and eventually everywhere. They were not always successful, but a whole Earth emerged and has survived thus far.

Conditions believed to be similar to those **during the origin of life** still occur today occasionally for us to study. In the hot vents and underwater plumes from ridges in the oceans, where plates are spreading apart, hydrothermal fluids emerge from the hot lower layers[1]. Even some bacteria live now such as they were in the waters of the very early Earth, when simple organisms were growing and surviving in adaptive evolution. When Mount Saint Helens exploded in 1980, the blast brought sulfur and other poisonous chemicals at high temperatures into the atmosphere, wiping out contemporary life in nearby Spirit Lake. However, different microorganisms were soon feeding on the iron, ammonia, carbon monoxide and sulfur in the lake, similar to the conditions on the early Earth when there was not much oxygen available as yet.

Our present atmosphere, with the oxygen we are breathing, is **not the original one** made during the formation of the earth. Solar radiation swept the lightweight hydrogen and helium gases away. This is different for the outer planets, which have retained theirs, because the sun's radiation pressure is hardly felt. A terrestrial atmosphere came from the actions of volcanoes, impacts of comets and the outgassing of rocks. After that atmosphere had cooled to less than boiling at 100°C, the composition may have been primarily of CO_2 with some N_2, CH_4, H_2O and CO like the atmospheres of Venus and Mars are still now, having practically no free oxygen[1]. The transformation to the present atmosphere of mostly nitrogen and oxygen began as an **interplay** of this early atmosphere with the evolution of life by algae on Earth, 2,300 million years ago, 2.3×10^9. Let us look at that life and the discovery of its evolution.

11.2 The Discovery of Evolution

The evolution discovered in the 19th Century was not yet the one of the interstellar medium, or of the stars, let alone the one of the universe, but it was the one of living species. Had the earlier ones be known, they would surely have been added as the foundations for life on Earth. The universe was unknown at the time, the physics of stars and interstellar medium unheard of. The evolution of people was therefore the one that was discovered; it happened about 1840 and became a hotly debated controversial event. The discoverers were Charles Robert Darwin (1809–1882) and Alfred Russel Wallace (1823–1913). While others before them had written about evolution, as Darwin's grandfather had, it was the recognition of **natural selection** as the predominantly controlling feature that made them the true discoverers. Natural selection has long expert definitions, and a short one in our Glossary: "the waveforms of a new species and its environment interact and thereby affect each other".

Of the two discoverers, Charles Darwin is the better known because he devoted his life to it, and he lived in England at the center of the debate. He had grown up as a rather wealthy young man who had no great plans for his future when he was a student. He liked to hunt and hike the outdoors. He studied some medicine, and theology in which he even received a degree comparable to our Masters. A change came to him and over him when family connections brought an invitation to be an **unpaid naturalist** for a voyage around the world on the British Navy ship HMS "Beagle," which was to last nearly 5 years from 27th December 1831 until 2nd October 1836. There had been a precedent for his post, some sixty years earlier when Joseph Banks had collected flora and fauna samples on HMS "Endeavour" with the famous Captain James Cook[4].

Darwin went for an interview with Captain Robert FitzRoy who expected him to follow the Bible by finding species such as finches the same wherever their ship would sail, because they had all been created the same some **6,000 years** earlier according to the religious belief of the time. Darwin expected to become a minister upon return from this voyage and did not express dissent at that time.

His disagreements with the beliefs of Fitzroy and the world were soon to come and glaringly so, because he shared the Captain's cabin on this long voyage with big collections of fossils and samples. Right from the start, he found evidence that species adapt to the local environment. However, it would take him two decades to publish the controversial theory. In September 1835, after having sailed around South America, he reached the **Galápagos Islands** where he camped for about a month. Darwin was not favorably impressed with the rather barren volcanic islands, but he could not help becoming intrigued with most of the animals being completely tame, for they had no predators; he could stroll among the boobies sitting on their eggs, paying no attention to him whatsoever. In 1999, it was still the same: I walked within a meter of them with four grandchildren, and the boobies hardly even looked at us. As a first observation of the Galápagos, Darwin wrote[4]:

> It was confidently asserted, that the tortoises coming from different islands in the archipelago were slightly different in form; ... I was also informed that many of the islands possess trees and plants which do not occur on the others... Unfortunately, I was not aware of these facts till my collection was nearly complete... I therefore did not attempt to make a series of specimens from the separate islands.

He did however, notice environmental differences between the Galápagos and the South American mainland, namely that the same species had become quite different in appearance. If the species had not evolved, they would have been the same everywhere. If they had evolved, they could be as he observed.

Years later, when studying the specimens he had shipped to England, he became more familiar with the **variations within** the species. Finches had found different foods available on different islands and so their beaks varied from thick and strong ones for cracking the nuts and seeds on one island, to longer but weaker beaks for catching insects and feeding on the fruits and flowers on another island. Gradually the idea formed that the animals and plants had originally been the same but had changed in their own **isolation**; we have noticed the effects of isolation on people in Sec. 1.1. Darwin also began to test his ideas regarding the **origin of new species** - a harder problem and greater end step than that of the variation - with further observations and experiments. He made use of the experience of people who selectively bred pigeons for specific characteristics, and he then broadened his theory with extensive discussion of examples in nature. The above two types of finches had become **different species**; they could not mate and produce offspring any more.

His search for a general principle got a start from the book *Essays on Population* by the economist Thomas Robert **Malthus** (1766 - 1834), who had concluded that living species produce more offspring than can be sustained by the available food supply (but see Sec. 14.2 for a correction). Darwin built onto Malthus' thesis with the idea of a struggle for existence, which is won by those who happen to be best equipped to do so. In the course of the earth's history, the living conditions change, and these local changes further affect flora and fauna. Variations or even new species emerge that are better able to adjust to the special or changed conditions. New characteristics emerge randomly, **by chance**, and the ones that are favorable for survival propagate because the owners survive better than others do. Survival through Evolution – Evolution through Survival. The variation does not happen because it is useful - it survives after happening randomly because it is useful. Local isolation sees to it that the kindred will mate so that the newly developed characteristics propagate (an extreme case is, however, inbreeding, with deleterious effects). The survival is thus combined with, and depends on the effects of the environment - Darwin called this **natural selection** and spoke of descent with modification.

He wrote several books, mostly to document his main theme described in *The Origin of Species*[5]. His contribution was so special because he **carefully provided the evidence** against the prevailing belief of the church at the time in our origin being brought about by divine creation. He was the scientist who was going to announce concepts that were counter to creation, and he would not have an easy time with it. Mrs. Darwin did not agree, she had some contradictory passages removed from his autobiography after his death[6], and the debate was never to stop[7].

Darwin had suffered from seasickness, but was in great shape on the shores, able to be on horseback for long and arduous fieldtrips, and writing reports with enthusiasm[4]. For the decades that followed at home, he mentions sometimes that he was ill; although he does not describe the symptoms, it sounds to me like a condition of **depression** caused by too much stress for too long. Others refer to his fascination with the beetles, that he was bitten and thereby given an infectious disease; one of his own stories tells how he had found three different beetles at the same time, did not want to lose any so he kept one in his left hand, one in his right, and had plopped one into his mouth.

He was a devoted and meticulous scientist who wrote in **easy language** that a general readership follows, still today. He described problems with his theory as well as the arguments in its favor, thereby daring to assist his critics. He began a "first draft" of *The Origin of Species* in 1837 and completed it in 1844, but he then still called it merely a sketch. He continued working on that until 1858, when he became aware of similar work and conclusions by Wallace, and then the first edition of his work appeared quickly. Some of the key contributions may well have come from Wallace[8], but Darwin's are the most recognized, perhaps because he remained focused on the problem; his Origins Book was meticulously improved through successive editions, of which the last one is dated 1872.

In overview of evolution on the earth, it has a long inorganic start, a stage from which species appeared in the oceans at first, and then evolved everywhere into ever more complex forms. **Slight deviations** from average characteristics occur, and these are favored, or not, through natural selection: whatever fits the environment best will have the greatest chance of survival. Offspring from those with the best-fitting characteristic have a greater chance of having it too, and then they are likely to survive longer and to reproduce the characteristic.

As for people, if the average number of surviving children were 6, as it was then in western Europe, their number would increase by a factor of 6/2 for successive generations, a geometric progression (the steep curve is sketched in Fig. 13.2). The environment also changes when it is affected by life; overgrazing brings barrenness and dust storms. According to Darwin, any being that happens to be better suited to survive in the changing environment will do so. The abilities are **naturally selected**, while the less-improved forms of life are usually extinguished.

The struggle is still around us, even for improvement, if not for survival: finches fight even when there is enough food, and people compete for more of everything. It is noted however that animals and people may also band together to strive for a common goal. **Altruism** occurs, an unselfish devotion or even self-sacrifice for the welfare of the group. A bird or rabbit may give a warning signal even if it thereby exposes itself to that predator - volunteers march off to war. We shall illustrate in Sec. 16.3 how developed brains allow people to think and choose so that the evolution is no longer of the strongest but has become that of the thoughtful. We shall complete the early history first.

11.3 The Tools and Emergence of Life

Now for the beginnings of life, consider the period between about 3,800 and 600 million years ago; that is between the end of the heavy bombardment of comets and asteroids we discussed in Sec. 10.4 and the start of a great acceleration in evolution of Earth. The dust and rubble between the planets in the early solar system was gone, mostly swept up, and cleared away; **sunlight came through** so that the transition could happen towards photosynthesis, taking energy from photons. The good greenhouse of Sec. 1.4 was in place because there had been enough CO_2 in the atmosphere; the temperature was raised some 15 degrees Celsius above freezing, and that was a necessary condition for the development of life.

A balance arrived, such that the temperature could be self-regulating, somewhat like with a **thermostat**. When the temperature rises too much, more of the CO_2 interacts with the water to make H_2CO_3 that is readily absorbed into the rocks, and the problem of too much CO_2 disappears. However, the rising temperature also evaporates water and that brings more cloudy weather - increased cloudiness diminishes sunlight from entering, so it cools again. More water freezes out when the temperature drops too low, and the skies become transparent again for an increase in the solar warming.

Table 11.1 has the summary for the development of life on Earth[9]. Near its bottom we see long periods for the development of basic chemical structures; at first it was as described in Sec. 11.1 for the foundations of life. At about 4,500 million years ago the first emergence may have occurred of a variety of **extended molecules** for carrying information and soon the very earliest cells, but the heavy meteorite bombardment brought a delay and new start at 3,800 million years (My) ago, as we have seen.

Table 11.1. Summary of the Evolution of Life on Earth.

Geologic Period	Ago (10^6 y)	Development
Recent	0 - 6	People
Tertiary	6 - 65	Higher Animals
Cretaceous	65 - 225	Dinosaurs
Permian	225 - 600	Reptiles
Pre-Cambrian	600 - 3,800	Fungi
Lower Pre-Cambrian	3,800 - 4,500	Chemical Evolution

About 600 million years ago, the **Cambrian Explosion** began with life forms bursting forth in the oceans and later on the shifting lands. Figure 11.1 shows the tectonic plates 200 My ago; comparison of this chart with Fig. 1.6 gives an impression how our plates are moving. Note the Tethys Sea, which was important in Sec. 1.1.

Clay surfaces on the shores may have acted quite like the surfaces of interstellar grains did before, such that molecules, cells, and species could meet and combine into even more intricate forms. Table 11.1 demonstrates the **acceleration** in the complexity, especially after 70% of the present earth's surface emerged within the last 225 million years. As species become more complex, they do not take more time to build the greater complexity. The opposite is the case, and it is a striking effect: as species become more capable, they evolve faster. We see the accelerated development of advanced life forms in the most striking manner for the human species: it took humanity only some 6 million years to reach its present stage – compared with the evolution of simple species like the fungi types taking as long as 3,200 My. If one hour on the clock represents the whole history of the earth, our human presence would happen in only the last three seconds.

Fig. 11.1. A sketch of the earth 200 million years ago. The oil fields of the Middle East are now where the Tethys Sea was.

Let us step back in time again to look at the tools of life's evolution[9]. Reproduction might perhaps have had a forerunner at the DNA stage. A speculative idea is that by bumping among and against other DNAs, the strands may sometimes have come loose, exchanged, and penetrated like in a mating. The basics of that were established even earlier in nature, as we saw in fusion and ionization with their collisions, and in the losing and gaining of components.

Soon an even greater tool for splitting and mating evolved, **the cell** (L, cella, small room), with its principal characteristic that components split and double. The first were simple and small, without a nucleus, they are **prokaryotes** (from *pro-* before, plus the Greek word *karyotos* for nut or nucleus). The first prokaryotic cells survived in simple environments that had only water, ammonia, methane, and carbon dioxide. Oxygen began production by the interaction of the primitive life of prokaryotes with sunlight. The environments gradually became suitable for a more complex nucleus to evolve, and eventually the much more capable cell emerged, the

eukaryote (the word's root is from *eu-* well or true). Eukaryote cells have a prominent nucleus, while the prokaryotes are much smaller and simpler in structure; they are more primitive because they had come about earlier. The eukaryotes were to make evolution more complex, more capable, and faster. Both types are still with us in the life forms we will overview in the next section; bacteria, for instance, are prokaryotes.

The variety of cell functions is impressive, what with **waste handling** and making proteins from the amino acids. The latter originate from foods and channel to the ribosomes for **protein production**. In plants, chloroplasts are the "power plants" of the cell, collecting water and carbon dioxide to make **sugars** for nourishment, while the oxygen comes free and evacuates out into the atmosphere.

Beginning about 2,000 My ago, more complex cells came to bring sexual reproduction, as we will describe below. The geological record shows the end of the preparation period by about 1,000 My ago with the emergence of complex **multi-celled** life. That preparation had also brought the usage and production of oxygen and the change of our atmosphere from the one mentioned at the very end of Sec. 11.1 into the atmosphere that we know today. The stage was set for complex life and a rapid increase, the Cambrian Explosion.

The nuclei of the eukaryotes contain the genetic material, which carries the characteristics of each species. The root of the word **genetic** is the word gene (Greek, short for pangen; pan, all; genēs, born). We speak of **genetics**, which is a branch of biology that deals with the heredity and variation of organisms. A **gene** is a set of instructions, important for dictating the variability in life; they are located on chromosomes, and encoded in the sequences of the DNA[9].

The DNA in Fig. 11.2 is a helix shape of two strands of proteins spiraling around each other, held together by hydrogen bonding. Combinations of four molecules, only four, connect the two sides. They are "bases", the **a**denine, **t**hymine, **c**ytosine, **g**uanine that are also shown. The example of DNA sequencing of Fig. 11.2 may be spelled out as ATCGGCTAGCATTACCG. Groupings of three bases specify one **amino acid**, which is the building block of a **protein**. The length of and the variety in such sequences store the characteristics of the species. It is a system with simple additions as evolution proceeds. The complexity of life on earth, from the simplest organisms to human beings, encodes in intricate and enormous variety for combination of only the four molecules. The same components are used over-and-over, but the **great variety of combination** stores complex instructions as for reproduction. The comparison with bricks of a building comes to mind, where in the choice of brick combinations a door or a window may be encoded, and a large variety in shape and structure. However, while some of the buildings on our campus may have a million bricks, each human DNA has some 3.2×10^9 of the ATCG molecules - yes, 3.2 billion!

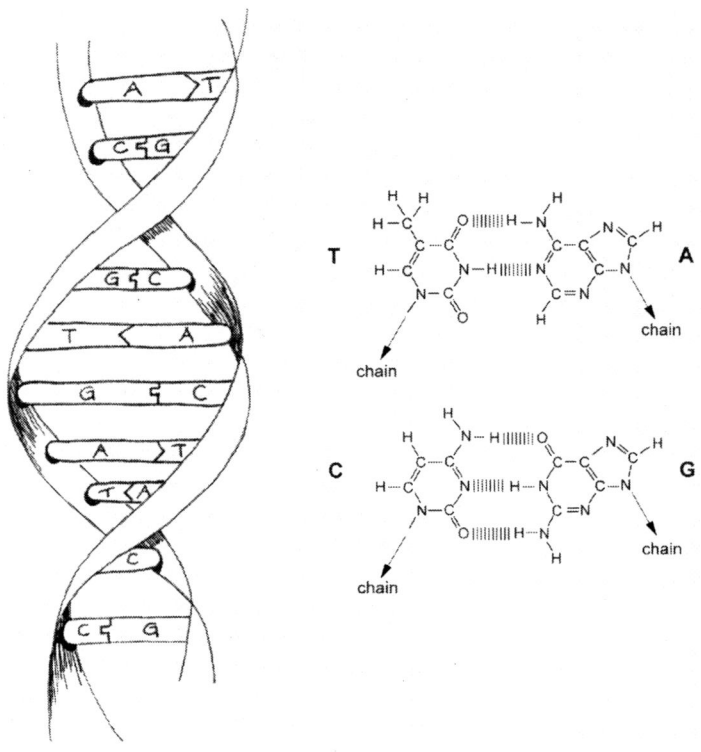

Fig. 11.2. The double helix of DNA, with its coiling sugar-phosphate chains. Thymine and Adenine are hydrogen-bonded, as are Cytosine and Guanine.

Astronomer Sig Kutter had the fabulous idea to compare the storage of information with the way we do it in a **library**[10]. An alphabet may have 26 letters – compared to 4 DNA bases. Letters are grouped in words – the 4 molecules in groups of three. Words, sentences, and paragraphs are comparable to increasing groupings to molecules, amino acids, and genes. Articles are arranged in volumes – thousands of genes in a DNA molecule. Volumes make up an A-Z encyclopedia – 23 DNA molecules, *i.e.* 23 chromosomes, make up the human genetic information. Each library is a housing for an encyclopedia – each human cell for pairs of 23 chromosomes, totaling 46. Kutter also pointed out to me that the sequence of pairs, the genetic code, is linear, one following the other, as are the sequences of letters in books and of the books on library shelves.

Chromosomes serve to package the DNA; they are threadlike structures of various sizes in the cell's nucleus and that is where the hereditary instructions reside. Each species has a specific number of chromosomes per cell. Humans have 46, that is 23 pairs, 22 being the same in men and women, but the 23rd pair, of the "X and Y sex chromosomes," determines the difference. During fertilization, half of each chromosome of the male combines with a half of the female's, completing the total again but now containing the characteristics of both parents. The combination of characteristics of two families is where we obtain the **variation within our species**.

A remarkable process occurs among the chromosomes, of **recombination**, whereby the characteristics of members of both families are scrambled. A baby may have the color of the eyes from one parent, of the hair from the other or from a grandparent or other forebear. That is what brings the opportunity to evolve from a wide range of characteristics, which strengthens the population; there is the U.S. example in Sec. 16.2. The sad opposite is **inbreeding**, when cousins marrying cousins, which brings deleterious defects together due to the close relationship of the parents, the lack of choice. Let us look at a selection of living species.

11.4 The Evolution and Variety of Organisms

In this Section, we go far back in time and then forward, in order to overview species from life's beginning until its present. There are books regarding the exact epoch of the beginning of life, and they have left much unanswered; another favorite topic is its definition, "What is Life?" The treatment in this book sees no exact or major epoch but rather a natural and gradual progression of evolution since the beginning described in Sec. 7.3, whereby evolution itself began in the multiverse.

Our journey in this Section will bring us along to various species, several names we are already familiar with, but if not, you may wish to memorize them, respectfully because these are our neighbors on Earth and **we depend on them**. At the same time, we also make an overview of evolution, which resulted in approximately the order of the text, and we will note a few of the most important attributes of the successive species. A background and overview of this section are in Fig. 11.3.

Most of the early species have died out, naturally, for example methane-producing **cyanobacteria**[11]. Without these prolific organisms, most of the life on Earth would not have emerged. They lived 2.3 billion years ago, before there was oxygen in the atmosphere.

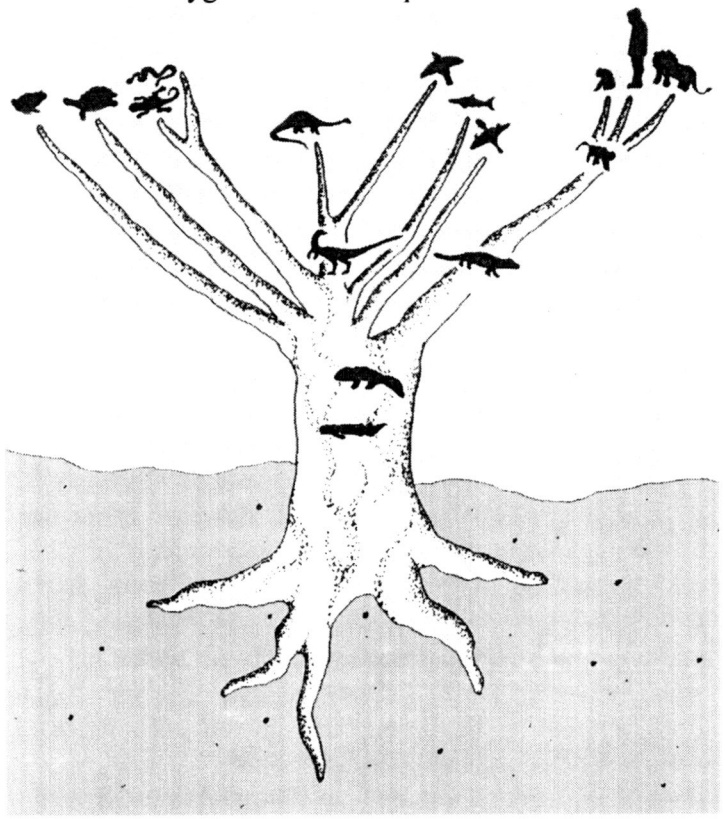

Fig. 11.3. The Tree of Life.

Of the early species that can be seen today, the very ancient and primitive **viruses** are regarded either as extremely complex molecules or as the simplest of microorganisms. They are not organisms that live alone, since they cannot reproduce themselves. These viruses are parasitic, living within and off host cells making new viruses according to genetic instructions provided by the virus. In this manner, they reproduce in large numbers, break destructively out of their host cell, and infect new ones.

In contrast to viruses, **bacteria** are living prokaryotes that reproduce by the division of single cells into two identical offspring. Bacteria frequently form colonies and are the most numerous organisms on Earth;

some of them cause disease, but the human death from bacterial plagues, that raged during the Middle Ages, have been reduced through improved hygiene and sanitation. Not all bacteria are harmful - some are important in the production of foods such as yogurt, cheese, and vinegar, and we carry bacteria to aid our digestion. In fact, if the bacteria disappeared from Earth, living things including plants and animals would soon die for lack of the ammonia and nitrates, produced by bacteria, that higher plants need for protein synthesis.

The structure of **amoeba** is simple, but we see the advance in evolution because they do already contain a nucleus and other intracellular membranes, they are eukaryotes and are essential for the origin of plants, animals, and fungi. They occur in various environments where there is water: in the seas or fresh water, moist soil, and the bodies of animals and people. An eyespot may occur - already this early in the evolutionary process - sensitive to differences in direction and brightness of light; the sun was of course an important part of the environment controlling the natural selection.

Fungi differ from green plants in that they do not yet have chlorophyll for photosynthesis, so they need dead or living organic material for their existence. Examples of fungi are yeast, bread and fruit molds, mushrooms, but also penicillium, a group of filamentous fungi used primarily to produce penicillin antibiotics. **Lichens** are the associations between fungi and photosynthetic algae. They are important food producers in both cold and desert areas, which may help to feed animals, or affect the formation of soil that can allow higher plants to grow there.

There is **plankton** in the seas and oceans, which serves as food for animal life and produces a large part of the world's oxygen supply. The small animal-like organisms in this group may be free-living, or they may be parasitic. Several of them bring diseases, and there are some highly poisonous organisms in this category; "red tides" of **algae** for example occur all over the earth and cause the death of numerous fish. The word "algae" is used for various kinds. The tiny green algae were probably the precursors of the land plants, and none of these should be confused with the less advanced "blue-green algae".

Land offers **plants** the advantages over water of more sunlight, oxygen, and carbon dioxide. The plants may then also have to deal with evaporation of the water, or with drought, and generally with lack of support for their structures. So they evolved with roots to obtain water and minerals from the soil, and with vascular tissue for transporting the water up, supporting the stems and leaves in the air, and for bringing photosynthetic food, carbohydrates down to various parts of the plant.

By this time, the **chlorophyll**, the green of plants, had established itself for the **photosynthesis** conversion of sunlight into food. Leaves had evolved, with chlorophyll inside, in order to expose a large surface area to sunlight. A picture of a landscape in infrared light shows any and all plants, even dead ones, bright because the chlorophyll has high reflectivity at longer wavelengths to protect the plant against overheating. Advanced plants further evolved and produced flowers to permit sexual reproduction. Plant structures and reproduction spread into a great variety of habitats on land; they evolved to be less dependent on moisture in the environment. Among the plant groups today, the algae are the most primitive, while flowering plants are the most advanced and best adapted to various environments.

Animals are distinguished from plants in that their cells do not have rigid walls, because their principal way of nutrition is ingestion and digestion of food, whereas most plants depend on photosynthesis for their nutrition. The diet of animals must contain a variety of substances such as fats, carbohydrates, proteins, vitamins, and minerals. By the process of the digestion, animals break the food down into molecules that can be absorbed through the blood stream into all parts of the body.

Among the simplest forms of animal life are the **sponges**. **Jellyfish, coral,** and **sea anemones** are already more complex and they generally have tentacles around the mouth for capturing their prey. **Flatworms** such as tapeworms are parasites and show special adaptation to that way of life; they can be meters long, deadly to people and must come out of their digestive system. The **leeches** and the common **roundworm** have a long cylindrical body divided into segments with fluid-filled cavities; these animals can be larger and more active than the earlier primitive ones. Leeches are familiar to soldiers serving in jungle terrain where it takes a little training to just let them be, have their fill of blood, so that they will drop off and not leave infections.

The **mollusks** are dominant organisms in marine environments, but they also occur in fresh water. Now we are dealing with rather complex animals such as the **squid** and **octopus**; the **snail, clam, oyster,** and **scallop** have shells. Some snails obtained evolved lungs when they adapted to land environments.

There still is an animal kingdom that is invertebrate, being prickly, but not having a backbone yet. They include **starfish, sea urchins, polyps,** and **sea cucumbers**; they are meat-eating sea animals, predominantly bottom dwellers for easy support. **Polyps** and their deposit are a part of **coral,** and the coral are a sad story, because they are vanishing due to over-fishing, pollution, and global warming coming from the lands. But let us stay for now with the new features that always appear in the progression of evolution - at the stage of this paragraph, they are the openings for the mouth and anus. Thus it may not be too surprising, in spite of the appearance of this group, that it is the most closely related to the chordates that we will see below, which includes humans.

We are getting to the arthropods (Gk, arariskein, to fit, and pod, foot): **crabs, lobsters, spiders, centipedes,** and **insects**, that have evolved in a wider range of environments and with a greater number of species than all other animal groups combined. A protective skeleton covers their bodies; jointed legs are another evolved feature. Insects have three basic body parts: head, thorax (L., thorax, breastplate), and abdomen. They have three pairs of legs attached to the thorax, and a single pair of antennae on the head, for touch, because their compound eyes are not sensitive enough at close range. Winged insects have either one or two pairs of wings connected to the thorax. With all that equipment, they are among the most successful of land animals, rivaled only by humans and other mammals.

The vertebrates (from Latin vertebratus, jointed, having a spinal column) consisting of **fishes, amphibians, reptiles, birds** and **mammals,** are **chordates** (L., chorda, cord), a large subgroup of vertebrates now with a nervous system. This listing shows that sea animals would emerge from the water, hesitatingly at first moving part time onto the land, where some of the plants had also come from the sea and these would gradually evolve too. At some stage, the chordata developed a flexible rod to which muscles attach, with a hollow nerve cord. Paired gill slits became successful as the left and right sides of vertebrates are symmetrical, and they have paired eyes and ears.

The earliest fish found in the fossil record did not have jaws, but from these animals came the **sharks, rays, skates** and the **bony fish** of today. Some of them evolved with paired fin appendages, and with legs and lungs for the amphibians, which were among the first animals to adapt to life on land.

Birds and **mammals** were to come from the reptiles, which evolved the ability to lay eggs on land with shells. Birds have a constant body temperature, an efficient circulatory system, and a thick coat of feathers that provides heat insulation. Mammals maintain a constant body temperature with a coat of hair, and they have sweat glands, milk-producing glands, and a four-chambered heart, while the embryo develops in a placental uterine environment. The species proceeded to become even more advanced

11.5 Further Advanced Life-Forms

Animals of greater complexity came later, such as rodent-size mammals and **dinosaurs** some 200 million years ago; they had evolved from reptilian ancestors.

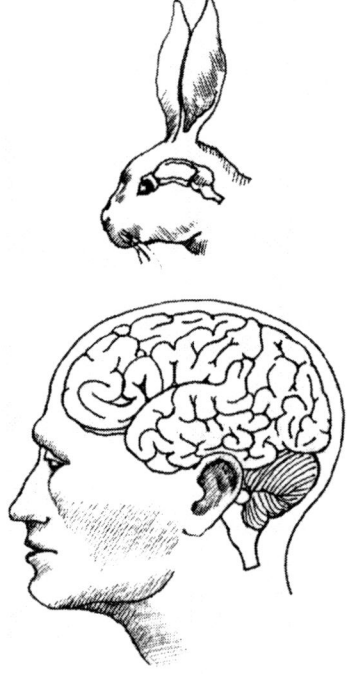

Fig. 11.4. Comparison of a little and a much developed brain.

A gradual evolution occurred, occasionally punctuated by ice ages and disasters such as asteroid impacts. An impact of a big comet or asteroid would cause the extinction of the **dinosaurs** 65 million years ago. It thereby became possible for the small mammals to develop into bigger ones, including people, because the big eaters were not around to consume them anymore and they now had some of the food available that the dinosaurs used to consume (Sec. 12.3).

We see a strong evolution of the **brain** in the animal kingdom, people included. Figure 11.4 shows a cross section through a simple brain and through the most advanced one. Near the brain stem lies the most primitive and basic parts, which control instinctive impulses such as the "fight or flight" reaction to danger; this brain stem emerged at least 400 My ago.

Immediately above it is a reception center for messages from all senses (except smell), and to command reactions to these senses. Behind the brain stem is the cerebellum, which coordinates the muscles of the body. The middle parts of the brain evolved for reptiles and primitive mammals to make basic survival possible; it now includes the instinct for parental care, which evolved between 300 and 100 My ago. Finally, the **cerebral cortex** followed, with a pair of lateral prefrontal cortices above the outer edge of each eyebrow, which allow memory, learning, and abstract thinking; differences between female and male are even found[11]. Speech added to growls and body language; the refinement of communication for how we socialize while having a meal together would take longer to develop.

Primates appear in the line of mammals to which humans belong. Good eyesight, agility, curiosity, and the ability to handle objects contributed to their success. The first primate forms evolved immediately following the asteroid impact of 65 My ago, but their fossils are scarce in tropical environments that have acid soils which tend to decay the bones. By about 30 My ago evidence is found of small monkeys and primitive apes. The Proconsul of Fig 11.5 fits about in here. Some 12 My ago, the great apes and, later yet, basic forms of humans emerged, with small canine teeth, leading to a reduction in face size, they may have used tools, were bipedal and sometimes walked upright. It appears likely that apes and hominids branched off from a common ancestor during that time.

Chimpanzees are closest related to humans, having DNA nearly 97% the same, with larger brains, sharp stereoscopic vision, a more upright posture than other primates and a tendency towards forward shifted foreheads. They have an omnivorous diet of plant and animal food, both, which leads to smaller and less specialized teeth systems; they can climb trees and have evolved a touch and grasping ability.

The great experiment of **upright walk**[12] occurred some five million years ago near the East African Rift Valley, which runs along the Awash River of Ethiopia, then the eastern shore of Lake Turkana, the small lakes south of Nairobi, and further south towards Lake Malawi. That is where two tectonic plates are presently spreading apart, but at that time they were coming together to cause some uplift, of merely a kilometer or two but that was enough to have the local winds rising to cause rain, an "orographic" effect, which resulted in a lush terrain with trees and growth for life to flourish. The Rift area has now become a dry region, but rich for finding evidence such as simple tools and skeletons of our human forebears. Walking on all fours, "knuckle walk," was already common, with even an occasional walk on hind legs as the bears can do, and dogs can be trained to do. Then our branch separated from that of the chimpanzees with changes in the bone structure so that it became more comfortable to walk upright. The higher vegetation of the trees encouraged them to be upright, while the increased choice of facilities, hunting and eventually travel, challenged the brain so that it would develop faster and faster over the next millions of years.

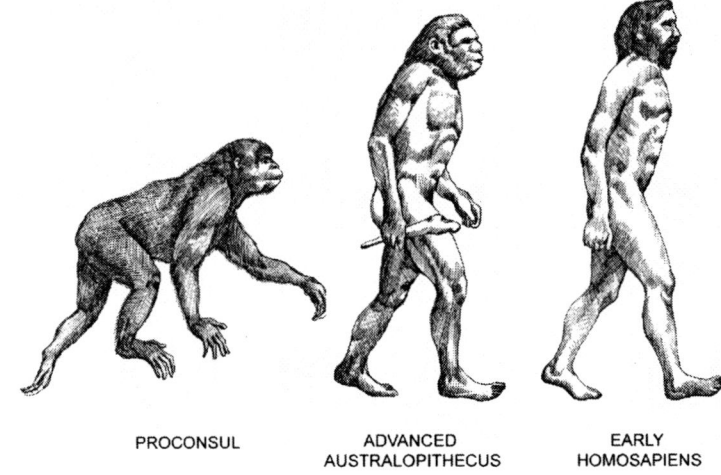

PROCONSUL　　ADVANCED　　EARLY
　　　　　AUSTRALOPITHECUS　HOMOSAPIENS

Fig. 11.5. Evolution through three recent stages.

Our ancestors may have had to adapt to new, open environments, and they started exploring for better places to settle into. The earliest **hominid** (a primate of the family Hominidae, of which Homo sapiens [L., wise man] is the only extant species) is Australopithecus, who stood up straight on two legs, had a large brain and perhaps tool-making ability. A more advanced form of **Australopithecus** that is in Fig. 11.5 appears to have used crude pebble tools. The traits characterizing both proto-humans and humans fall into four categories: increased brain size, fewer but more generalized teeth, erect posture to walk on two legs, while sexual reproduction with care for the young had come earlier but it now became dominant. All of these features were present also in a side branch Homo Erectus (L., upright man), which is known to have used tools for hunting, they made fire and built huts for shelter, and they engaged in cannibalism. Their remains of 500,000-years ago are on Java and near Beijing.

Homo Sapiens probably also came from Africa (Fig. 11.5); a 160,000 years' old skull was found in what is now Ethiopia. There was side branch that became extinct some 30,000 years ago, the Neanderthal species, with evidence found first in the Neanderthal, a valley (thal) near the present city of Düsseldorf, Germany[13]. They hunted mammoth and other animals, gathered edible plants, and probably lived in caves, in single-family or small multi-family groupings. Specialized tools were used, adaptation to a variety of environments increased, their migrations began some 2 million years ago and eventually spread to nearly all parts of the earth[14]. They hunted and were hunted at first, but they were better equipped and gained the upper hand, learning to make fire and becoming evermore adept at using tools and ships, and at living in larger communities with increasing communication and interdependence. Each gender adapted to different roles within the community to fulfill different purposes, all aiding the survival of the group.

Major but gradual transformations occurred in the ways people harnessed matter and energy and made themselves more powerful, namely fire control and agrarian and industrial revolutions[15]; to that mix I would now add globalization. **Fire Control** began about 1.5 million years ago, and the widespread **agrarian** ways of living some 10,000 years ago when people had already spread all over the world but gradually needed more food than could easily be found or hunted[14]. They became less tied to the land, began to live closer together, and their tasks were divided and delegated into the various trades. The usage of gunpowder began in the Middle Ages and that brought an equalizing in the ranking of people, and eventually the end of the feudal society. We see city and state formation a few thousands of years ago for greater protection and delegation of tasks, but with that came increased exploitation by a few powerful of many others, tax collection, and military control, up to the present times. The **Industrial Revolution** occurred in the 19th Century, its social excesses well known from the books written by Charles Dickens (1812-1870). The **globalization** of trade, industry, and communication is a modern stage that we shall discuss in Sec. 13.3.

As much as we understand evolutionary processes, human evolution must be **continuing**, but the time span of the human species may be too short for observing great effects. European castles have armored uniforms on display, but men in that part of the world do not fit in them. There is evidence that the appendix is on its way out as a useful organ, we can do without the spleen, gall bladder, or one kidney. People's environment is evolving noticeably with changes in society, business, and industry. By natural selection, the species will evolve too[16].

Great discontinuities in universal evolution have occurred in the distant past: when particles formed, when the electrons combined with nuclei to make atoms, when atoms combined into molecules, and with the emergence of reproduction and the variation of species. What other **discontinuities** will evolution produce next? It seems impossible to make predictions, but one could think that biotechnology and electronic communication are among the leading candidates for making drastic changes[17]. With electronic brain chips one can make deliberate changes[18]. The rapid increase in usage of the Internet may change society into a much better educated one.

The question facing us at the beginning of the 21st Century is how far genetics, robotics, nanotechnology, and information technology might affect us and cause an appreciable change in our evolutionary development. As for genetics of animals and plants, with DNA instructions taken from the one into the other, species can be duplicated, cloned. Cloning of plants has been going on for a long time, but strong objections protest doing this kind of experimentation when animals and especially people are involved.

For the effects that electronic communication may have upon our evolution, I have an example that would interesting to investigate. It concerns the word-processor generation, with its members having been at computer screens since early childhood. That must have instilled an exact **obedience**, a conditioning in obeying commands, because the slightest error in the typing brings non-action or failure. Has this generation become more obedient and less likely to protest and remonstrate as the young generation of the 1960s did against war and regimentation.

Part III. MODERN SOCIETY

12. NATURAL DEATH

12.1 Multiverse and Galaxies

The travels through space and time have brought us to the last part of this book, which concerns the future and we will discuss the ultimate destiny first; the word 'destiny' is used because it looks certain that our universe, galaxy, solar system and we people also are unavoidably aging towards death. The word **predicament** is used for that concept in the sense that the only certainty we have in life is that we will die. We may actually learn something from discussing first the other deaths - of universe, galaxy, and especially the sun - in order to put our own in a broader perspective and gain insight and acceptance of our own death as an integral part of universal evolution.

In this section, the stability of the multiverse is mentioned first because that seems fundamental, and it certainly is on the largest of our scales. The exponential relation in Sec. 6.6, $M(\alpha) = (hc/G)^\alpha$ indicates infinity because there is no limit for its α exponent. The $M(\alpha)$ model appears to imply that it is **without an ending and without a beginning,** because there seems to be no scientific way to stop it, or to have started it all at once, or to have started it increasingly. Each epoch has its own limitations in the sciences, and we must leave the issue there. Scientists, however, will never leave such a tempting problem of origins out of their future.

The theoretical models of our universe being a pulsating one (Sec. 7.2), seem to have vanished [except one in string theory (Sec. 7.3)]; in fact, there is an acceleration of the universe's expansion. That by itself would not constitute an ending other than for extragalactic astronomy; the intergalactic space expands, while the internal distances within each galaxy do not increase. However, the universe ages and it decays because its galaxies do.

The compositions within galaxies show an **evolution sequence** whereby the irregular galaxies came first, and there are only a few percent left among our galaxies, next there are spirals, while the elliptical galaxies, with their old stars and little gas and dust left, are in the majority. Our galaxy is already beyond middle age.

We also observe collisions among galaxies[1]. Their stars mostly pass by each other because there is so much space between them, but their orbits may be gravitationally perturbed. **Collision effects** come from the clouds of gas and dust, which have large cross-sections such that they collide with an increase in the formation of new stars. In the end, after a few of such encounters, the two may become closely associated, their gas and dust mostly gone, and they may retire together into an elliptical galaxy with old dying stars as the main inhabitants.

Another way to look at the aging is how we displayed our galaxy evolving from being young with hydrogen and helium only, to getting gradually older with the predominance of heavier elements. The formation of galaxies was at a maximum rate 5×10^9 years ago (at age 8.7×10^9 for the universe), and the stars within them are gradually running towards a decreasing supply of hydrogen and helium fuel. The speed of transition by the stars towards their demise depends on their original mass, and the end-products fall in the order of white dwarfs, neutron stars, or black holes (Sec. 5.4). Our **universe is slowly dying with aging stars and galaxies** as ours. The dying of universes is a basic process in the multiverse, as its universes continually **recycle**; the debris of radiation, atomic particles, and other mass are used in accretion of new masses for new universes. However, the dying of our universe does not affect the lifetime of our species, because the timescales are much longer than that of the sun, which determines the end of life on Earth, as we shall see.

Our own Milky Way galaxy may have a specific problem. The **Andromeda Nebula**, our nearest spiral-galaxy neighbor M31, is coming our way, as Slipher discovered by finding its blueshift, indicating its approach (Sec. 6.5). However, the distance of M31 is about 2.9×10^6 lightyears and the velocity towards us is only 121 km/sec = 4×10^{-4} lightyear per year so that the collision occurs $2.9 \times 10^6 / 4 \times 10^{-4} = 7 \times 10^9$ years from now, long after life has vanished on Earth. Besides, it is unlikely that the currently predicted trajectory of M31 is so well known that a collision is assured, and the two may pass by each other with little effect other than perhaps on our Oort Cloud whereby a storm of comets may be unleashed.

Another galaxy is seen in the Sagittarius Constellation, and it is colliding with ours at this time, but it is a small one. It is being absorbed without any noticeable effect on Earth. We do live in a world of collisions, among asteroids as well galaxies, and a special surprise is yet to come.

12.2 The Solar Neighborhood

The chance of collisions by stars with the solar system seems negligibly small, because the mean distance between stars is several lightyears as Fig. 4.7 shows[1]. The solar system does occasionally get relatively close to interstellar clouds of molecules and dust, which are larger than our entire system but they are tenuous. The frequency of such a collision is about once in ten million years or so, but I have never heard of any detected effect. **Gravitational tides** caused by close encounters to the Oort Cloud by stars and molecular clouds have been postulated for bringing storms of comets towards the inner solar system, as we mentioned at the end of the M31 story.

However, at a public session of the International Astronomical Union in 2000, an astronomer showed in alarming detail in three dimensions a **star coming at us!** It is star number 710 in a catalog made by astronomer Wilhelm Gliese, so it is called "Gliese 710", passing through the Oort Cloud a million years from now. It is not clear what effects we should expect from Gliese 710 other than horrendous comet disturbances and impacts on Earth. Furthermore, the prediction so far in advance must be quite uncertain and will need to be checked as time progresses.

Events that could harm our atmosphere could come from **supernovae** if they come to some 10s of lightyears distance, a rather unlikely event[2]. **Gamma-ray bursts** are dangerous if they would occur anywhere on our side of the galactic center (on the left in Fig. 6.2). They are likely to have made a number of significant impacts during the last billion years. There may have been enough gamma radiation to cause intense ionizing in the upper atmosphere, breaking up molecules of oxygen and nitrogen and destroying the ozone, which would have severe effects on life. Complex effects on the terrestrial atmosphere may also have caused events of global cooling[3]. The hazard we see next is precisely modeled, namely the sun's variation in brightness.

12.3 Death of the Sun

Graduate astronomy courses have been teaching the modeling of physics and action in the interior of the sun and other stars since the 1930s. At various layers, the students compute the pressure and temperature of a gas under gravitational attraction and obtain the nuclear transformations and energy production from astrophysics. The demise of the star begins when the available fuel supply has burned out at the center[4]. Energy generation will still occur in shells farther outward, and the star as a whole will then expand because the heat source shifts to shells that are larger, at greater radius from the center. However, the resulting overall heat generation may not be enough to keep the large outer surface of the sun white-hot, so the sun would be observed from the outside with a reddish color. The expansion and the color change together will give it the appearance of a **red giant** (Sec. 5.3). Later, after the fusion supply begins to be depleted in all of the central layers, the star will enter a period of **instability** with erratic expansion followed by shrinking, pulsation, and change in overall output and brightness. With the instability and variation in brightness, some material may shed outwards into a shell around some of the stars in a spectacular display[4].

The sun will ultimately die in a highly concentrated stage of burned-out matter, of nuclei and electrons not bound anymore in atomic configuration, but collapsed together. It is a **degenerate** configuration of a **white dwarf**, collapsed material in high density, which may last a billion years. There will be no nuclear energy generation any more in those years, only a slow cooling during which the amount of emitted light will be diminishing. Eventually our abode fades into darkness forever.

The models for such gaseous spheres give the following numbers. The sun is presently about halfway through its life, so it has a total of at least 5×10^9 years still left. However, already after **a billion years or two**, the size of the sun may expand by about 5% and its total output of energy may increase by about 15%; this will cause a temperature increase of about 10 degrees centigrade (10° C) on Earth. Such effects will be serious, for instance in melting of snow and ice, and in partial evaporation of the water on Earth.

About 4×10^9 years from now, the sun will be deeply in the above red-giant stage; the solar diameter will have increased by about 30% and its total energy output by 70%. The earth's average temperature may have risen by 50°C; water and life will have long vanished. As a red giant star, the sun's expanding radius may reach as far as the orbit of Mars. The sun's material of charged particles will drag the earth's orbiting to a halt - the **earth becomes a part of the sun**. Eventually the unstable sun will explode away most of its material.

In summary of these sections we conclude that **no terminal effect on life** is seen for at least the next billion years as far as the sun is concerned, let alone the galaxy and universe; demise through the supernova and gamma-ray effects is always possible but has low probability. There is, however, a natural danger that may eliminate human society long before a billion years from now.

12.4 Hazards due to Comets and Asteroids

There is a chancy hazard to life on Earth due to impacting comets and asteroids that is however within our power to diminish[5,6]. The objects that helped build the planets are not completely used-up so that some remain or are supplied anew - the contrast of the asteroids' character is sometimes referred to as being both **building blocks and wrecking balls.**

The first two columns of Table 12.1 show approximate numbers of dangerous objects that occasionally come close to the earth, described as **near-Earth objects**. It is seen that the numbers in the second column increase by approximately a factor of 100 for each factor of 10 decrease in size; the numbers increase rapidly for smaller objects. It is debris from collisions having such an increase, and that effect can even be seen when a brick is hit with a hammer because there are many resulting small pieces. We are dealing here with debris from collisions in the asteroid belt, flung around the solar system by Jupiter's gravity.

Table 12.1. Approximate Numbers for the Asteroid Hazard.

Diameter (km)	Number of Objects	Approximate Number of Years between Impacts	Impact Energy (in "Hiroshimas")
10	11	5×10^7	10^9
1	1,100	5×10^5	10^6
0.1	110,000	5×10^3	10^3

The third column gives the estimated frequency of impact on Earth[6], but it is an average because there is nothing predictable about individual events: the chance for the collision to happen today is as great as for any day in the past or future. The last column gives the energy of impact, the well-known expression for kinetic energy $E = 1/2\ mv^2$. We will see that asteroid impacts are horrendous because the mass, **m**, and the velocity, **v**, are large quantities, and the latter is squared (a driver-license question is usually asked of how much worse an accident is at twice higher speed, and the answer is 4 times, not 2). The velocity is the relative speed of near-Earth asteroids with respect to the earth's surface, which can be as much as 100,000 kilometers per hour. For Table 12.1, the mass is determined from the diameter in the first column converted to a volume, with the density of meteorites adopted at 4 grams per cubic centimeter, which is a compromise taking the most dangerous metallic ones into account.

We are dealing with frightening energies. To give you some idea of that, I have expressed them in terms of the energy expended by the **Hiroshima atomic bomb** explosion in 1945. This is not callousness about human suffering, but a feeling for the magnitude of impact disasters because we can relate to the horror at Hiroshima and Nagasaki (in Sec. 13.1 we shall discuss the effects of nuclear radiation, which the asteroid impacts do not have). In military and physics language, the Hiroshima bomb had a power equivalent to that of 12,500 tons of TNT, tri-nitro-toluene explosive, which amounts to 5×10^{20} ergs (a unit of energy).

The impact of an object with a diameter of one kilometer would carry the energy of a million Hiroshimas, equivalent to that of **all nuclear weapons** the U.S. and Soviet Union may have had ready for launch during the Cold War. This is in a single explosion, while the effects of the weapons would be spread over a large area through the impacts of many bombs. Asteroids of 1 km and larger are assumed to cause destruction on a global scale and the death of millions, perhaps billions of people. The probability for anyone of us to witness or perish by such a collision is large in view of the consequences. By dividing the years of a lifetime, say 75, into the 500,000 of the third column, it is one-in-7,000 in a lifetime, and that was the probability when we began surveying them in 1988. But now, more than 800 of the 1,100 have already been surveyed and found to be harmless at this time, so that there are about 300 uncertain ones left, thus the probability is 1,100/300 x 7,000, about **one-in-25,000** in a lifetime at present (the number has an uncertainty of several thousand). It seems a low probability, but it is for a calamitous event.

The potential suffering caused by an asteroid impact is so terrible that it would be irresponsible *not* to take care of this problem. It may be **the most serious hazard** humanity faces because the impact by a 1-km asteroid can obliterate global society in one blow; severe disruption of our international activities and communications would occur even for smaller impactors. The encouraging point of the asteroid hazard is that searches diminish it, as the change from the above 7,000 to 25,000 shows. Half a dozen telescopes - in Arizona, New Mexico, California and Hawaii – are searching with the goal of finding the ones having a diameter of 1 km or larger. Spacewatch, the program of the University of Arizona, has two telescopes on Kitt Peak, one for surveying, and the other for follow-up on the discoveries. Anticipating further developments in facilities, 90% of the 1,100 in Table 12.1 will be known and have their danger assessed by about 2015; the probability of an unexpected impact will then be one-in-70,000. The hazard will however never be completely gone because new objects keep coming from the asteroid belt and from the Oort Cloud[5]. Objects smaller than 1 km cause smaller impacts, but crashing into an ocean they can still cause widespread and destructive **tsunami** waves. The U.S. Congress has indeed recommended, and NASA has accepted, to continue searching of the skies for smaller but still dangerous objects.

It is essential to follow-up newly discovered objects that appear to be hazardous, in order to know more precisely when they might hit. The media occasionally announce that a dangerous asteroid is on its way, but up to now, all these reports have been wrong. An object at first **appears** hazardous, but always becomes less so through additional follow up.

A longer observational interval yields a more precisely known orbit, *i.e.,* **a smaller predicted-target area,** so that the cross section of Earth invariably is no longer be included.

Number 99942 in the Asteroid Catalogue deserves a special mention in this connection. It is a 0.3-km asteroid called Apophis, the Destroyer, and that naming was done because it appeared at first likely to hit the earth on April 13, 2036. However, by observations over a longer time (also made with radar so that the distance was determined as well as the position on the sky), the chance of a collision became nearly negligible. Radar observations are due again in 2013 and, if danger is then certain, there will probably be enough time to stage **mitigation through complex rocketry**[6]. There is, however, a group of scientists, astronauts and others promoting more aggressive action against the impact hazards[6].

As for the mitigation, in case a collision with the earth would appear to be on its way, the alarm would be for diverting the object's trajectory such that the impact is avoided[6]. One of the techniques proposed for that was by exploding a chemical or nuclear charge near the asteroid so that the blast would cause a deviation of the trajectory large enough, accruing over time, to avoid impact on Earth. Advance warning and lead times of decades are required for the accruing because of the enormous mass and velocity of the asteroid - this is **different from immediate destruction** of an incoming projectile as is done in warfare with much less massive objects. The charge would explode before the rocket would reach the asteroid, in a **stand off** explosion, in order not to destroy the asteroid; its fragments would cause more damage than the single body would, as is known in hunting when one uses buckshot to cause the greatest impact. There are estimates of the power needed for delivery near the asteroid and for deflection of its orbit, and they seem feasible using the nuclear and rocket capabilities of the U.S. and Russia. It would be ironic but of deep philosophical satisfaction if the dreaded nukes could save humanity after all! There may be, however, other means of changing the trajectory of a menacing asteroid, such as by having an ion-propulsion engine on its surface (Sec. 14.7), and this may be preferred because it is gentler and may not break up a loosely configured asteroid[6].

Among planetary scientists, there is no doubt of the need to identify the possibly dangerous objects as soon as possible. It is not a question **if** a sizable asteroid could hit us but **when**. However, the public and government authorities still need to become convinced as to the reality of the problem. Because it is such a theoretical hazard, of low probability and witnessed by no person known to us, it may be useful to discuss some examples of events that have already occurred.

A classical case of a sizable meteorite impact is the one of the "**Meteor Crater**" near Winslow, Arizona (a misnomer since meteors are small pebbles). About 50,000 years ago, a metallic object about 60 meters in diameter made an impact crater 174 meters deep and 1.2 kilometers in diameter. More recently, in 1908, a brittle object of about 60 meters in diameter caused a tremendous explosion in the atmosphere, perhaps making a small crater, near the Tunguska River in **Siberia**; people as far away as London heard its rumble.

The most spectacular example of an impact in the solar system occurred in 1994 by more than 20 comets into the atmosphere of Jupiter[7], each having diameters near 1 km. Table 12.1 shows that if the impacts had been on Earth, it would have been a catastrophic calamity. Actually, a collision worse than that because of its effects occurred on Earth 65 million years ago, by an asteroid or comet about 10 km in diameter that caused the extinction of nearly 60% of all species. Its impact site is a crater of some 180 km in diameter near where presently is the town of Chicxulub in Mexico's Yucatan Peninsula. The event caused wildfires over huge areas of the earth from fiery pieces coming off the object due to friction with the terrestrial atmosphere. Next, a large dust cloud blew out from the explosive impact into the atmosphere, so thick that sunlight did not penetrate for perhaps half a year. This prevented photosynthesis and growth of green plants as well as the weather-making capability of the sun. The dust cloud eventually settled in a 1-2 cm layer of geological deposits marking the end of Cretaceous time, while the layering above is Tertiary; it is therefore the **Cretaceous-Tertiary Extinction**. The composition of the thin layer is largely non-terrestrial, containing relatively high concentrations of metals. When the earth was molten during its bombardments by planetesimals, heavy metals such as iridium and platinum became rare at the surface because they had sunk towards the deeper interior in differentiation. Iridium and platinum occur in that 1-2 cm layer as abundantly as in some meteorites, and that discovery opened the discipline in 1980 for studying and avoiding asteroid impacts.

There still may be some debate whether or not the single impact of the Chicxulub event killed the dinosaurs. The doubt comes particularly from paleontologists, who study fossils and point out that the dinosaurs did not disappear suddenly. What probably happened was a **global warming** of about 15 degrees centigrade (compared to our present fears of a few degrees), caused by the original fires, liberated CO_2 and other gases from deep deposits of an ancient tropical forest there. The warming was to last thousands of years and that may have done the dinosaurs in. Other mass extinctions such as happened 251 million years ago have been interpreted with asteroid impact, but they may have been due to global warming followed by heat and H_2S gases emerging from the earth and sea. In any case, the dinosaurs eventually were not around anymore to eat the mammals, which were then among the species that could survive and evolve unimpeded by the dinosaurs. The mass extinction of 65 million years ago therefore had far-reaching **consequences for the propagation** of biological evolution. We may be here now because of the demise of the dinosaurs.

Human society will, however most likely be **eliminated by an asteroid or comet,** long before the 10^9-years decline of the sun. New objects keep coming from the asteroid belt and from the Oort Cloud. There also may be comets and asteroids with warning times that are too short for defensive action. That will depend on the circumstances, particularly how convincing the discovery is (time to convince governments of the reality), combined with time needed for the applied orbital-deviation impulse to be effective, which depends on its size and configuration.

Meteorites have not yet killed any human being as far as we know. In 1957, in Alabama, a meteorite went through a roof and a ceiling, bounced off some furniture, and then hit Annie, who was resting on the sofa, on the hip[8]. The **story, or fable,** is that Annie wanted to keep the meteorite as a souvenir, but the owner of the rented house wanted to sell it to pay for the damage to roof and ceiling. Then the State of Alabama entered into the case because this was obviously an Act of God while the State represents the Highest Authority. The State proposed that the precious sample from the heavens be stored in a museum for all to see. A judge had to decide this case, and he assigned the meteorite to Annie because "she had the most intimate familiarity with it." Years later, I telephoned to verify this story, but Annie had passed away and her divorced husband hung up on me because he did not want to talk to yet another reporter.

12.5 Death and Rejuvenation

Now that we have seen evolution, aging and demise everywhere and at all times, it may be easier to accept our own **predicament that we must die**. The previous four Sections were to give some perspective on our own death, how demise is prevalent in the universe. The Sections show that death is a universal predicament - everything in the organic and inorganic forms of planets, stars, and galaxies seems to fade away. It may be impressive, however, that it all may used again in the recycling of universes in the multiverse.

Let us include the history of our components into this philosophical view of demise, death, and rebirth. The Big Bang or multiverse origins of our universe are not mere theoretical exercises in astronomy. Every one of us was involved. In the multiverse, and in the early phases of our universe, all of our subatomic parts were present; they produced every component for our 10^{28} atoms, yours, and mine. Each one of our atoms went **its own way** as it was flung out into different regions of multiverse and our universe. They passed through various stars and interstellar grains for the formation of molecules. Each of our molecules then had its own history during the violent formation of the earth. They passed through the water of the oceans and the gases of the atmosphere, through plants and rocks, and through other people.

How many of our atoms were in famous people like Jesus Christ, or Mohammed, George Washington or Mahatma Gandhi? I do not know how to derive that rigorously, but a **million** or so (10^6 of the 10^{28} atoms within us) seems a reasonable number. Of course, the infamous ones, like Adolf Hitler, are included as well. An estimated 10^{10} people went before us, 10^{16} of all their atoms thus being within every one of us. I do not imply anything here from the realm of fantasy because the atoms are just what they are, without a signature of, or a message from, where they resided before – they are perfect atoms but without the tools for each to deliver a message. These 10^{16} atoms are scattered throughout our bodies and they are not even permanently installed as they come and go; they do not combine to make a voice or memory. The use, the only use, of this paragraph is to emphasize - and to celebrate delightfully and respectfully - how interrelated we are in our matter and in our origins.

Inside ourselves there is also a continually changing population of molecules as we daily take large numbers of them in and put large numbers out. The molecules within us come and go with our food, breath and other actions and with the wear and tear and aging of our bodies. We consume about 2,500 calories per day, which is the energy rate of a **110-watt light**

bulb. We see a continuous renewal occurring when we are alive, and we may speak of a re-incarnation afterwards in the sense of participating in new forms. Parents are keenly aware of that and can look with love and pleasure upon their children continuing the family, passing on the genes that give us grandma's beautiful hair, etc.

Here lies the broadest of connections. Our atoms continue in other people and aspects of nature. We ourselves consist of atoms that have been in others and elsewhere, and that is the beauty of our being an integral part of our world and the cosmos. Each one of our molecules from our breath goes on to live for a long time in inorganic or organic forms, each on its own path into the future. After death, the bonds of complex forms such as cells, neurons, and chromosomes are broken down, bones persist the longest in a burial site but not forever, not even in mummy or mausoleum preservation. Atomic bonds are stronger, but even they are loosened during the aging and demise of our universe, and the mass seems preserved and a new start is made through quantum fluctuations and new atoms and molecules following in other universes. When we die, whatever molecules are left will go on into other organic or inorganic forms, back into the air, soil, water, plants, animals and people. From a grave, crematorium, or funeral pyre, the atoms and some molecules of the body emerge onto dispersed paths, largely into the atmosphere. This is the ultimate example of **rejuvenation** for the components of planets, stars, galaxies, and universes and, perhaps in a few places here and there, in their oceans, plants, and intelligent species. The traveling molecules, atoms, or fluctuations connect you and me with the animals and environment, with this universe and other universes, in their past, present and future.

There are still other concepts that may help the understanding and **acceptance** of death. While sub-atomic particles seem to have a nearly infinite life, their more complex combinations and the results of evolution have much shorter lifetimes, from perhaps 10^{50} years for the proton and less for more complex atomic nuclei, to 10^7 year for O and B stars and 10^{10} years for later type stars, down to 10^2 years for people. This last sentence indicates that we must die because greater complexity is more difficult to maintain. Let us work that concept out carefully.

The intriguing aspect of the evolutions we have seen is that they are part of the same **universal evolution**, which began in the multiverse. Evolution evolved itself into ever more complex entities with greater information content and therefore more capable techniques for subsequent species. So it happened that the subatomic particles evolved, followed by the atoms and then molecules, each in their own different environment, while the evolution on Earth brought a development of molecular strings, of which the most advanced was DNA, which originated reproduction as an advanced evolutionary tool.

However, the mechanisms of evolution had become so complicated that their products were weaker and **more vulnerable** to destruction. Worn out components are rejected by the body, replenishment of molecules from the outside is needed - heat, food and drink. An intricate factory for doing all of that had developed, the cell, with its variety of functions. It had evolved new capabilities, but also greater need for replacement of worn-out components and the necessity of rest and sleep, of decay and death. After that price was built in, the cell could become so powerful that evolution was enhanced rapidly, until it took only a few million years to evolve the presently most complex species. These species are feeding and excreting actively, while they have also become vulnerable to tiredness, sickness, aging, and death. **Life had brought its own demise.** We are paying for our capabilities with having to nurture life and to live a short life. The life expectancy for people was originally only 30 years or less; the Tara Umara Indians in Northern Mexico still live only some 35 years. The wealthier people will soon be extending their life expectancy towards a hundred years, but that is still vastly less than the lifetime of the proton of at least 10^{35} years.

Death may be easier to accept in view of the fact that everything seems to die, including the stars and the galaxies, our universe. Death may be easier to accept from our understanding of universal evolution, that death is the price we pay for a fully lived life, and for having our offspring as a part of the renewal and rejuvenation processes. We may ultimately gain the Upanishads' insight,

"When the eye is free of death it becomes the sun."

13. MAN-MADE PROBLEMS AND SOLUTIONS

13.1 Nuclear War and Radiation Effects

Hiroshima was destroyed on August 6, 1945, by an **atom bomb**, which exploded by fission of the heavy element uranium; as Japan did not immediately capitulate, Nagasaki was next, three days later with a plutonium bomb, the only other atom bomb ready to be used at that time. These nuclei are unstable and can be made to split, releasing enormous amounts of energy in the form of a blast of gamma rays, x-rays, other high-energy particles, neutrons, and optical radiation in a blinding flash of light.

A visit to the **Atom Bomb Museum** in Hiroshima is haunting even to hardened veterans[1]. Near **Ground Zero**, immediately below the bomb's explosion, only the lucky ones were killed by the blast and flying debris, while most of them died in radiation agony lasting as long as 24 hours with their skin and flesh slowly drooping from limbs and body. The indescribable horror and its extent to thousands of people come over in wax figures and actual photographs. The number of people who died at that time in Hiroshima was probably more than 60,000, but the total in the following 5 years may have been more than 130,000, primarily from radiation effects. **Fall-out** from the atmosphere had also occurred, of radioactive particles that cause further damage to flesh and bone[1,2].

The question whether or not the United States **should have used** the atom bombs in 1945 on the two cities is still being debated[3]. I remember as military in the field, that some of us condemned the usage, even while knowing that it might save our lives, because the argument for destroying German cities had been that they had used the technique first, at Warsaw, Rotterdam and British cities; using atomic weapons, with their radiation fall out, would justify any others to do the same.

On October 28, 1962, however, the U.S. and U.S.S.R. came close to nuclear war in the Cuban Missile Crisis[4]. One could make the argument that the demonstrations of the effects in 1945 had persuaded everyone against starting a nuclear World War III, and that this argument may apply still today against a **first strike** with a nuclear weapon. While the chance of a global nuclear war has greatly diminished since the end of the Cold War, the danger of using nuclear weaponry in warfare or terrorism has not diminished at all. We live in an aggressive world, in which some 40 conflicts may be going on at the same time[5], some of which large enough to merit that the United Nations has about 15 peacekeeping missions, while the trade and manufacturing of new weapons seems unbridled.

The original atomic bombs had a limitation in the range of sizes; larger ones would explode by themselves, while for smaller ones it would be difficult to ignite the explosion. Bombs became more powerful soon, making helium from tritium, which is hydrogen that has a nucleus with one proton and two neutrons. Such weapons have multiple layers – if it is spherical, there would be concentric shells of various materials; centuries ago, oriental style fireworks already had a shell system for their spectacular bangs. Today, a bomb may have an inner core of uranium or plutonium like the above atom bombs, around that core a jacket of chemical explosive charges to cause the uranium or plutonium to fission, and an outer jacket with the tritium material. Without further jacketing, it is a **neutron bomb** because it releases many fast neutrons; they are for localized usage because they kill people more than cause damage to structures. For greater explosive effects in this gruesome business, another booster jacket fuses nuclei of various hydrogen isotopes towards the formation of helium nuclei in order to form a **hydrogen bomb**.

By the end of the Cold War in the late 1980s, the U.S. and the Soviet Union probably had about 10,000 strategic warheads each, with a total energy of perhaps as much as 500,000 times that used at Hiroshima[2]. The strategy of the time was that of **deterrence**, preventing an attack by relying on the common sense that no one would initiate a nuclear war because it would result in **mutually assured destruction (MAD),** and the U.S. and USSR both made assurances that they would not strike first.

Since that time, there has been considerable **reduction following negotiation**, but there still are frightful weapons in various countries[5]. And then, in 2002, the U.S. Congress approved the development of a new nuclear weapon, the Robust Nuclear Earth Penetrator, the so-called "bunker buster," for the penetration of mountainous caves[6]; after protest, pointing out that radioactive fall out would kill innocent people, the funding was withdrawn, perhaps temporarily. There also were initiatives for developing other new types of nuclear weapons, a new production complex for building them, and funding for decreasing the amount of time it would take to resume nuclear testing. Another initiative followed, of the Reliable Replacement Warhead, and proposals for replacing all nuclear warheads.

Both the U.S. and Russia continue to patrol submarines with large nuclear rockets at presumably undetectable locations, thousands of nuclear weapons are still poised for immediate launch, and large numbers of small nuclear charges are available for firing from artillery, airplanes, and cruise missiles. **Nuclear weapons** are also in China, Great Britain, France, Israel, India, Pakistan, North Korea, and perhaps secretly in other countries. Opposite opinion to this outcry due to these weapons being so debilitating by the effects of nuclear radiation, might be that their design and fabrication stimulate industry and employment, while the world has learned to live with a nuclear standoff to avoid usage by anyone, a policy that has been successful, with no nuclear weapon used since 1945, not even small tactical ones.

There is concern that a launch might occur accidentally or by a rogue nation or launch crew. The U.S. is therefore working on a system called **National Missile Defense** (NMD), with the aim of shooting down any missile coming in, as the American Patriot did to Iraqi rockets in 1992, but now with a nuclear warhead. Recommendations by the Union of Concerned Scientists and by the Center for Defense Information (CDI, which is guided by senior military officers) are speaking out against nuclear weapons because of the radiation and fall-out hazards, NMD violation of existing international agreements, and because the renewed re-armament may bring about an **international arms race** again[6,7]. Russia and China have noted that one can use an NMD system also offensively, and Russia has warned that it may have no choice but to include NMD sites among their potential nuclear targets. An American plan to build silos for launching NMD rockets in Poland has particularly aggravated Russia, as happened in reverse during the Cuba crisis. The U.S. has openly announced that it feels no longer bound to any no-first-strike assurance because of first strikes by Osama bin Laden and his **Al Qaeda** organization, starting in 2001 and continuing today.

We also face a peacetime problem of nuclear pollution from energy generators. We have seen how stars generate energy by means of nuclear **fusion**; atomic nuclei grow into larger helium, carbon, and other nuclei. In Sec. 14.2 we shall mention how the search for man-made fusion, as happens in the stars, is conducted in laboratories, which is a great hope for the future.

In sharp contrast are our manmade reactors, which generate much less energy from nuclear **fission**, which is the breaking up of nuclei that are so large that they are unstable and therefore always radioactive; the heaviest nuclei used are those of uranium. The Hiroshima and Nagasaki bombs had the same fission characteristics. The fission leaves us with still-large nuclei that are highly unstable and keep sending out radiation in the form of subatomic particles with great speeds and thus great energy, and we therefore have dangerous **radioactive wastes**.

The hazard of the wastes is measured in terms of their **half-life**, the length of time it takes for half the atoms of a certain batch to disintegrate, and that can be as long as several thousand years. Disposal sites must be leak proof for such lengths of time during which the radioactive nuclei decay and the energetic particles continue to be dangerous to life. It is difficult to find such sites; the U.S. government assigned a major one in Nevada in 2002, over strong opposition, because some evidence of underground leakage in futures of thousands of years had been indicated[8]. Various countries dump wastes in the ocean, as is done for city garbage, even though radioactive materials can then enter the bodies of the fish and thereby of people.

There have been two infamous **accidents** of nuclear-energy reactors getting out of control. The worst was at Chernobyl, north of Kiev in the Ukraine with deaths of crew and in the neighborhood, while fall-out from that 1986 accident drifted as far north as Sweden, where two decades later there still was an abnormally large number of people diagnosed to have cancer. During a near-accident in 1979, the crew narrowly averted an explosion at the Three-Mile Island reactor near Harrisburg, Pennsylvania.

In spite of these concerns, new nuclear generators are being built in the U.S. and elsewhere. Environmentalists criticize that as a ploy of and for wealthy industry while cheaper solutions are ignored. The opposite view is that the accidents are avoidable, and that a solution may be found for the storage of waste products, perhaps by re-processing, and the next paragraph has an example of that.

A connected but little-known type of hazard, in-between war and peace, is in **radiation effects from spent munitions**. For many weapons as of the 1980s, U.S. forces rely on depleted uranium, which, being nearly twice as dense as lead, can penetrate defensive armaments such as for tanks more effectively than conventional alloys can. The metal is a by-product of uranium enrichment for nuclear power plants and warheads, but it is toxic when ingested and always slightly radioactive; there are United Nations' attempts to clean up after wars[8]. We have to proceed now to examine even

13.2 Chemical and Biological Warfare

In this section, I will have to paint a dismal picture for the menace of chemical and biological warfare, but this is the situation, as I understand it. A limitation of these weapons is that they work well only in confined spaces, or ventilation will dilute their effects. Anyway, the major lesson and reaction should be for us to remain informed by the Union of Concerned Scientists and the Center for Defense Information[7], and to write to representatives in government when there is an opportunity towards control or elimination of the hazards; such letters are also welcome from outside the U.S.

We begin with **chemical** warfare that uses **inorganic** molecules. The usage of gas in World War I was at first with chlorine and mustard gases - the word mustard meaning an irritation - during the warfare of 1917-18 in Flanders and northern France where the two sides were deadlocked in their trenches. The French were the first to try to break out of their trenches by using gas, but that was tear gas, which is non-toxic, still regularly available to police forces. The Germans however released a chlorine gas that killed 5,000 French soldiers. The other 10,000 in the trenches fled their posts in panic. The Germans were actually so surprised by the effect that they did not follow the fleeing French and make a breakthrough. Later in that war, both sides used gas occasionally. Many of the survivors were maimed for life with destruction of their lung cells, which was called "having **gas lungs**". Accidents still happen when people are careless and breathe in chlorine fumes when chlorinating a swimming pool.

Following that war, most nations signed a Protocol, except the United States, **banning chemical and biological warfare,** but not the production of such weapons. This Protocol appears to have worked remarkably well because during World War II, gas was not used at the fronts, even though both sides had large stockpiles, and both sides got into desperate situations. There may have been technical reasons for not using gas, such as that the prevailing winds might have caused one's own troops or a friendly population to suffer. However, in confined locations such as in special barracks of German concentration camps, millions of people perished with gas.

Another international agreement, signed in 1975, outlawed **germ warfare**, and **biological** warfare fought with **organic** molecules or in **germ warfare**. Attacks with chemical or biological weapons occurred in Iraq in the 1970s and early 1980s in a war with neighboring Iran, and on Iraqi minority populations[9]. The United States used chemical weapons during the Vietnam War to deforest hiding places for guerillas and enemy troops. Poisonous substances affected civilians, and the defoliant known as "Agent Orange" has affected American soldiers too. Over the decades, the U.S. has acknowledged the existence of large amounts of stored chemical ammunitions[10].

The state of biological warfare is shrouded in secrecy; we simply do not know what kinds of weapons exist and in what quantities. Bio-warfare could be the **weapon of choice;** humanity may be on the brink of frightful conflict with biological weapons[11]. The U.S. announced in 2002 that it will use all available counter measures to a biological or chemical attack, including the use of nuclear weapons. Let us inspect a broader background of the various forces and stresses in the world.

13.3 President Eisenhower's Warning

Regarding the controlling powers in the world, former U.S. Republican President Dwight D. Eisenhower made a surprising and unforgettable television address[12]:

> In the councils of government we must guard
> against the acquisition of unwarranted influence,
> whether it is sought or unsought,
> by the military-industrial complex.
> The potential for the disastrous rise of misplaced power exists,
> and will persist.
> We must never let the weight of this combination
> endanger our liberties or democratic processes.
> We should take nothing for granted.
> Only an alert and knowledgeable citizenry
> can compel the proper meshing
> of the huge industrial and military machinery of defense
> with our peaceful methods and goals,
> so that security and liberty may prosper together.

Eisenhower had gained broad experience as Supreme Commander of Allied Forces in Europe during WWII, as President of Columbia University, and as President of the U.S.; he spoke deliberately on the momentous occasion of his **farewell** television address in 1961, with millions watching and reports reaching a world-wide audience[12].

Instead of diminishing the military-industrial complex, the warning became more of an omen of what was to happen, such that the number of globe-spanning conglomerates increased rapidly[13]. The effects occur in many places, for instance where owners **leave small farms** or businesses because they find it impossible to compete with the economies of feedlots and supermarkets. On the side of the large businesses, it is felt that their economies make the products available at lower price for the consumer; their stockholders, who are not necessarily wealthy people, cannot help but agree with the productivity.

Consolidated businesses of the U.S. and other industrial nations are spreading over national and international trade, and their **lobbyists** are in the hallways of the people's representatives in U.S. government. Alarming is the extent of limited-access and tax-deducted entertainment hosted by industry at both the Democratic and Republican Conventions. Eisenhower also learned when he was President of Columbia University how for scientists a government contract becomes virtually a substitute for intellectual curiosity, and he warned how public policy could become the captive of a scientific-technological elite. The establishments and businesses are richly rewarded, sometimes by government involvement in the form of preferential tax policies; this has boosted some of the rich to become multi-billionaires, and there are shocking reports of cronyism and corruption[14].

The businesses have expanded abroad wherever a profit is likely; **globalization** of industry is making it worldwide in scope and application. Eisenhower warned that the industrial complex tends to control or at least influence its business environment. There is information regarding **worldwide fear of the U.S.**, as *Newsweek*[14] explained, "What worries people around the world above all else is living in a world shaped and dominated by one country - the United States of America. And they have come to be deeply suspicious and fearful of us." My view is that it is not just the U.S. but more generally big industry headquartered in many countries not even limited to the West, and also that the people do gain in their standard of living - I have seen that happening in India. But I am concerned when the profit motive begins to dominate over everything else, for that makes it a harshly calculating world.

What of the future? Who will ultimately be in charge, nationwide and worldwide, and what will be the motives and ethics of the people in control? Will we then have to summarize the situation with our definition of **fundamentalism** for the leaders having "a point of view marked by adherence to the profit motive as its basic principle?" Or, can the leaders be persuaded that they can afford to mind the future beyond the profit motive and follow humanitarian principles such as sponsoring the poor as well? The record shows that the answer may be "yes" to some extent because all large companies set funds aside for philanthropic contributions; education and expression of public interest in the companies may help to promote such philanthropy. We shall see an example of coaching poor children in Sec. 16.7, with the participation of personnel from major companies who meet with the children and guide them toward a higher level of education and employment.

Can individuals mitigate the decline in democracy? Eisenhower seems to have thought of that question so he warned that the **citizenry must be alert** and knowledgeable, taking nothing for granted. The first Prime Minister of India, Jawaharlal Nehru, gave an example of international calamity when the citizenry was not knowledgeable. When he was in British imprisonment for Independence activities, he wrote an overview of world history for his daughter Indira, who later became Prime Minister herself[15]. While the causes of World War II are clear, those of the war of 1914-18, WWI, are not; the problem is aggravated because the pertinent historical records have been destroyed[15]. The reasons for the disappearance together with the causes of the war may become clear from Nehru's analysis:

> ... how the greed of capitalistic industrial countries, the rivalries of imperialist Powers, clashed, and made conflict inevitable. How the leaders of industry in each of these countries wanted more and more opportunities and areas to exploit; how financiers wanted to make more money; how the makers of armaments wanted bigger profits. So... the youth of the nations rushed at each other's throats. The vast majority of these young men, and the common people of all the countries concerned, knew nothing of these causes which had led to the war.

We should consider the possibility that Nehru's warning could be applicable more generally in combination with Eisenhower's, because profit motives are always present. The German military-industrial complex supported the rise of Adolf Hitler. With modern expansion and with weaponry more available all over the world, the risks and consequences of warfare are much more rapid and serious. The future looms with an even greater threat because of the re-emergence of an international nuclear-arms race (Sec. 13.1). The opposite view is that chemical and biological weapons are easy to develop, that even nuclear ones can be produced by **menacing regimes**,

or isolated terrorist groups. Our world's citizenry seems little aware of or alarmed by the consequences of living in such a weaponized world, so let us study that from its history.

Up to the time of the attack on Pearl Harbor in December 1941, the United States had favored an **isolationist** position, but then the transition from isolationism to globalization had an explosive expansion due to the needs of allied warfare. In Europe and Asia, the military were soon to see the results of organization, production, and fighting, accomplished on large scales through the mobilization of industry, on both sides. After World War II, the GI Bill lifted veterans into higher education, and the **Marshall Plan** helped the European nations recover; they and the U.S. re-established trade relationships and became strong allies in the "Cold War" against the Soviet Union. There was deep fear in the western world of the rapid expansion of communist regimes in Europe and Asia. The Cold War boosted the industrial complex such that by the end of the 20th Century its worldwide expansion had become a **globalization of commerce, manufacturing, and communications.**

When the Cold War had ended in the early 1990s, a **peace dividend** seemed possible from the reduction of military needs and expenditures, expected because there was no worthwhile enemy left in sight. The people were going to have fascinating times again like in the 1960s when the **moon landing was the symbol of the time.** The American people did eventually pay off their huge national debt[16]. But the 'defense' budgets began to rise again, to the point of the combined military expenditure of the Western nations in 2002 being 5 times as much as that of the others in the world (Russia, China, and all other countries combined)[17]. Why was that done - how did it happen? It actually seemed quite natural at the time, how free enterprise found markets, people commonly became stockholders of the conglomerates, while the widely expanded interests and Americans abroad seemed to need protection. Economies were doing well nearly everywhere for all but the poor, in the expansion of weaponry and electronic communication markets, people were busy with their own pursuits through the affluent 1990s, and the **large automobile was the symbol of the time.**

For the citizenry to be knowledgeable, we should recall more in detail what happened in that affluent decade of the 1990s. I will combine my own notes with **a thoroughly documented book** by senior political scientist Chalmers Johnson[18]. Johnson published his analysis in the year 1999, based on his extensive experience especially in China and Japan. My own experience goes back to WWII and the Cold War[19], to interviewing people during my yearly journey to India and elsewhere, and by preparing for the teaching and the writing of this book. Whereas Johnson concentrated on the United States, I believe that the problems are broader, with commercial conglomerates growing all over the globe, such that the influence of large European and Asian industrial complexes should be included.

The military complex is, however, American, and after the Cold War, its bases abroad stayed, even when the locals wanted them out. The situation is different from the days of WWII and the decade after that when Americans and their military were welcome and felt proud to be American. The U.S. Central Intelligence Agency uses the name of **blowback** for the presently negative local reactions. Two extreme examples of blowback, that caused American withdrawal, occurred in Lebanon in 1983 when a suicidal car bomber killed 241 U.S. Marines, sailors and a soldier, and the withdrawal from Somalia after an American helicopter crashed and one of the bodies had been dragged through Mogadishu.

Commentator William Pfaff spotted an intricate case of possible blowback due to Western monetary transactions in Indonesia, which he saw[20] as "an episode in a reckless attempt to remake the world economy, with destructive cultural and social consequences that could prove as momentous as those of 19th-century colonialism"; a detailed analysis is in the last paragraph of 16.6. I visited my old stamping grounds in Irian Jaya in 1992, but was not allowed outside Jayapura by Javanese military because of Papua uprisings for Independence. Their opponents were Javanese who had been armed with U.S. support (there is vast copper mining in Irian Jaya), and the same was being reported in Aceh, Northern Sumatra, where American companies are pumping oil (American help for the 1994 tsunami was concentrated on Aceh). The opposite argument to all this is that all Representatives and Government officials, Republican or Democrat, have the responsibility to assure continued supply of oil and other resources.

Various authors generalize and yet punctuate the situation by referring to "The American Empire". As before, I broaden this to include the influence of large European and Asian firms, so that I would call it "**The Industrial Empire**". Anyway, the empire comes from the use of capital from a variety of capitalistic sources for bringing economic integration about on its own terms and on its protection by American military power everywhere[18]. The proponents of the empire believe that the globalization of policies is working well, bringing democracy and effective economies so that they work as successfully as in the developed nations[21]. Opponents of that point out that the trade and finance are now operating in different backgrounds and societies where that culture may not be understood or wanted; there are large areas where democracy may not be desired[18].

Chalmers Johnson documents eye-openers of little known and sometimes covered-up cases of empire and blowback in Cambodia, Guatemala, Turkey, Japan, Chile, Honduras, and Nicaragua. He understands **empire** as a modern form in history that lies concealed beneath concepts such as "alliance", and "the free world".

Ronald Steel, an analyst of international relations[22], had pointed out already in 1967 a **characteristic that is little known** or understood: "Unlike Rome, we have not exploited our empire. On the contrary, our empire has exploited us, making enormous drains on our resources and energies." For example, the United States exports the most advanced weaponry, then contends that more must be invested in arms development at home because it is necessary to keep ahead of the world, and the U.S. citizens are taxed to pay for development of the new weapons.

The industrial complex in the meantime brings goods to these U.S. citizens that is made in foreign sweatshops at prices with which industry at home cannot compete, such that there is a basic shift in manufacturing away from the U.S., which was in many cases the country of origin and invention[18]. The work is stopped at home, the company is closed, but it may be opened anew in a country with low standards of pay and healthcare. Or, the product may simply be manufactured in whatever foreign shop, and imported back home, cheap. This is called **outsourcing.**

Johnson's surprising discovery, and his fear, is that **arms sales by the government** have become one of the Pentagon's most important missions and that its industrial policy is creating potential problems that may prove beyond all solutions. Commercial firms are of course also selling arms to whoever want them, and sometimes the arms' industries themselves are sold with it. The U.S. has armed even **both sides** in conflict in the cases of Iran and Iraq, Greece and Turkey, Saudi Arabia and Israel, China and Taiwan[18,9].

Thoroughly undemocratic is selective release of the news, seldom in open lies but more by the withholding of truth and in subtle manipulation of media and opinion polls. There is the amazing effect on the human brain that something frequently repeated becomes the truth; this is of course the principle of **advertising**, which is penetrating and pervasive in world society. I find an amusing and yet challenging example when I see one of those huge billboards that merely states, "Milk is Good for You." Can we trust that, when the cows are injected daily with the Bovine Growth Hormone, and when they eat chicken manure and "protein concentrates" of fecal and cadaver origin, as we shall document below? More seriously, the main impediment to new resources, healthier living, environmental improvement, and safeguarding the future is by vested interests in the **status** quo. Eisenhower warned repeatedly for this aspect of our society in his words "guard," "unwarranted," "democratic processes," "take nothing for granted," and "alert and knowledgeable". We must keep informed and be inquisitive, it is not anti-American to do so - it lies in fact at the root of the Nation to safeguard its principles and enterprise.

One of the worst secrecy excesses reported by Chalmers Johnson is the penetration by the military's **Special Forces** in so many countries (110 by 1998)[18]. He gives examples that show violations of human rights and sometimes even direct presidential orders. Turkish Commandos received crucial training with which they reportedly killed some 22,000 Kurdish people in eastern Turkey[18]. Johnson has the perception that the U.S. has come to use only one means of achieving its external political objectives, namely by military force, and that we rarely see anymore the presence of or reliance on a seasoned, culturally and linguistically expert **diplomatic corps**. I will report similarly on the usage of military force *versus* **Interpol** effectiveness of intelligence penetration in the hunting for terrorists (Sec. 16.6). Johnson made the following interesting suggestions[18], in the year 2000:

1. Work with the United Nations as a respecting member.
2. Adjust to and support China's emergence on the global stage.
3. Recognize global diversity rather than globalize Western standards.
4. Extricate the military from most foreign bases.
5. Re-emphasize true "defense" for the Department of Defense.
6. Unilaterally reduce U.S. stockpiles of weapons of mass destruction.
7. Sign and ratify the landmine and international-court treaties.

I would add and put at the top, "restrict the manufacture and sales of weaponry everywhere;" and I will make a suggestion how to execute that in Sec. 16.7.

Any Secretary of Defense might react to all this by pointing out that he works for a President who is **elected by the people**, and that the Senators and Representatives, similarly elected, make the laws that guide the Nation. We might agree that this book must then serve to give observations that may help the voting people decide.

My own observations of the U.S. began in 1950 when I had an unexpected opportunity to hitchhike from coast to coast, and then decided to immigrate because I had found an efficient, self-assured and yet generous population. Even the drivers of expensive cars would stop for this hitchhiker. The drivers were open with the stranger they would never see again after their long ride together, so they told freely how they lived, how their government worked, and members of both political parties seemed not to sound extreme or fanatic, but rather tell jokes about the opposition[19]. Eisenhower's speech was a surprising warning of things to come. In case you think that the above eight suggestions are new or extreme, listen further to the Republican President's speech with which this Section began:

> Down the long lane of the history yet to be written,
> America knows that this world of ours, ever growing smaller,
> must avoid becoming a community of dreadful fear and hate
> and be, instead, a proud confederation of mutual trust and respect....
> Together we must learn how to compose differences not with arms,
> but with intellect and decent purpose.

13.4 The Global Warming Emergency

Internationally together we live in a **biosphere** (Greek bios = life), the part of the earth in which life forms occur, which includes the terrestrial atmosphere, its oceans and its mantle below us. The words **ecosphere** (Gk. oikos = house) and **ecology** are also used to remind us that we have reached the point where the previously dominant environment is now largely under our control and husbandry. Its calamitous result thus far is summarized with the authority of the National Geographic Society[23]:

> Humanity is squandering the capital bequeathed by billions of years of evolution. At a rate that is breathtakingly swift by geological standards, we are depleting minerals in the ground, nutrients in the soil, and ozone in the stratosphere. We are drawing down an irreplaceable reservoir of genetic diversity..

This gives a hopeless feeling that we cannot go back, cannot restore much of the damage done. We can do our best. I will try an approach of mentioning the main problems and possibilities for interfacing with the environment properly, and for doing our best to halt the damage, while reporting opposite views as before. Before the 1960s, there was little awareness of protection and conservation.

In Arizona, we let the Glen Canyon Dam for hydro-electric energy generation be built in the Colorado River and its Lake Powell be named after a famous explorer and conservationist, but then we rose in protest when the next dam some 150 km farther downstream was proposed, and that was won easily. The 1960s were a remarkable decade worldwide, and the **environmental awareness** was one of its results. In reaction to years of neglect, the year 2006 brought a sudden increase in such awareness again, which brings the hope we need. We must recognize that people and their needs are a part of the environment, and that the **big damage cannot be restored** - there is no way to bring back the unique beauty below Lake Powell that was there in towering rocks' shaping and landscaping seen for example in the 1965 film "The Greatest Story Ever Told". The proponents of the dam had pointed to the fact there are plenty of red rocks left along the Colorado River, that there would be more people boating than there were hikers, and that our prosperous standards need the energy.

Ecologists speak of degrees of increasing environmental action with the adjectives of **conciliatory, preservative**, and **restorative** conservation. The latter may even include doing something new such as tree planting in India's city of Ahmedabad, which is becoming noticeably greener. Other examples are the progress reported in birth control because the basic problem of pollution is the increase in the number of polluters (Sec. 13.5), while Section 14.2 mentions how a factor of as much as 60 in food requirements can be saved.

A key problem in these issues lies with pollution by industry, generally perceiving that the profits would go down with environmental action, which may be a mistaken opinion[23]. A valid question is, "Who should **pay** for cleaning the environment?" "The one who fouled it," is the obvious answer, but the open pits near old and new mining towns in Arizona indicate that the spoilers have been allowed to walk away from the mess. At the other extreme, the citizenry has sometimes forced a remarkably good clean up, such as in London and Pittsburgh where the atmosphere was badly polluted before the 1960s.

We should insert at this point the issue of the many **landmines** left in various places on the globe as an environmental problem[24]. Suggestion 7 of Chalmers Johnson refers to them (Sec. 13.3). Great Britain's Princess Diana was famous for her involvement and effectiveness in removal of mines and for helping the victims. The mines are from previous conflicts, while others are new for ongoing warfare. An international Mine Ban Treaty was signed by 142 countries (but not by the U.S., Russia and a few others), prohibiting the use and even production of these devious devices that maim or kill some 15,000 people each year. The latest news from the U.S. is that they are back in production, wholesale.

Regarding **pesticides**, the Institute for Food and Development Policy publishes evidence regarding their use and misuse. Large companies in the United States are exporting certain pesticides banned or restricted in the U.S. itself. By subsequent import of agricultural products, the dangerous pesticides are spread over the world and do come back to the U.S. The usage of pesticides is however declining thanks to genetic engineering in agriculture, especially in the U.S., whereby the products themselves are made pest resistant.

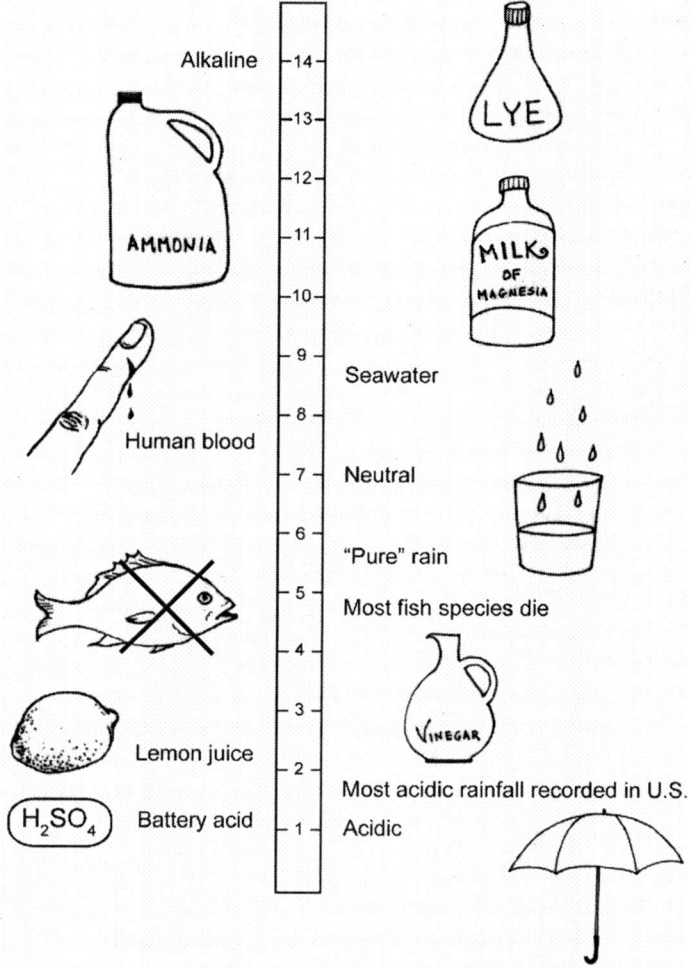

Fig. 13.1. The acidity indicator of pH numbers, illustrated by what the numbers mean.

The name **acid rain** is used when the acidity of the rains is excessively high as it is caused by volcanoes or industrial air pollution. Figure 13.1 shows acidity in terms of pH which has a peculiar definition as the *p*otential for *H*ydrogen, with pH = 7 for zero acidity and decreasing in number for increasing acidity. Acid rain happens when the atmosphere is polluted with toxic acids. Sulfur dioxide (SO_2), for instance, combines with water vapor in clouds or lungs to form sulfuric acid, H_2SO_4, which destroys even tough substances like leather. Acid pollution is becoming a global problem because the pollutants are carried large distances in the winds away from the original sources. Much of the CO_2 from the burning of fossil fuels goes into the oceans where it changes the acid balance of seawater at great dangers to marine life[24].

The prevailing wind directions in southwestern U.S. are from west to east, and sulfur dioxide produced by our **copper mines** may end up as acid rain over large distances eastward; as for large distances, there are reports of Beijing's pollution being recognized at the west coast of Canada. Anyway, the SO_2 from our smelters (that presently makes sulfuric acid with the water in our lungs) could be used instead for commercial production of H_2SO_4, but this is not as profitable as simply letting it fly into the atmosphere. The control regulations have in the meantime been side stepped by building the smoke stacks as high as 250 feet, and that has been exposed as a profit scandal[24]. We see these tall stacks now, after we had seen members of Greenpeace demonstrating from high up on the older ones. Furthermore, the pollution now comes mostly from northern Mexico; the Arizona copper industry has largely moved and triumphed there for lower salaries, health and pollution standards, *i.e.*, higher profitability. Some of the copper industry is coming back to Arizona, with modern automation, and now sometimes has difficulty finding the properly qualified personnel. The opposing side of these criticisms aimed at outsourcing, points at the increasing demands for copper and the advantages of having it produced cheaper.

Major sources of sulfur-oxide pollution are **coal-burning power plants**, which are responsible for about two-thirds of the SO_2 and one-third of the carbon dioxide (CO_2) emissions. These amounts in the earth's atmosphere have increased noticeably after a century of burning fossil fuels like coal, oil and gas. Recent increases have come from automobiles, which bring carbon dioxide, methane, and nitrous oxide. Its peculiar odor can be smelled only when the sense is refreshed in sleep in a clean climate – in Arizona, I am shocked by the first car going by when I am on the bike early in the morning, because it is an **awful odor**, but I cannot smell the second car anymore. Most people will never be able to smell this, or there might be much greater concern – what does this do to young children's lungs?

Many of these effects cause an excess of **global warming**; I call the excess caused by all these sources the **bad greenhouse effect** (the good one is in Sec. 1.4). There is always hope for the future such as a possibility of pumping CO_2 underground[24]. On the other side of the issue is a **denial of the problem**[25]; sometimes it is a devious political denial. It may be in the interest of the industry to deny the reality of some of the problems; one may be inclined to accept the denial because of the complexity of the problem. Academicians have confused the issue by writing scholarly papers about the difficulty of and uncertainties in measuring the warming trends because there are large variations in weather's temperatures. There is a dubious environmentalist's book, praised by the denial side, condemned by the conservation side[26]. In the U.S., this has indeed become a political problem, but we are discussing it here as a scientific problem.

There are large variations in climates caused by small temperature variations, or anomalies, in the Pacific Ocean called **El Niño** and **La Niña**. The temperature of the water mass may be only a degree or two different from normal, but it is over large areas of the ocean and can therefore have a great effect on the temperature of the air above, and thereby on the weather. The name of "El Niño" came from the Spanish words for "The Child" as these anomalies in rainfall occur around Christmas; the name La Niña is for an opposite effect that is smaller and caused by other temperature patches in the Pacific.

What may happen if global warming is real? Predictions are made with **meteorological modeling** that the amount of CO_2 in our atmosphere will be doubling in the next 30 years, largely coming from automobiles, and that this will raise the average atmospheric temperature by 4°C. A 4-degree rise will cause catastrophic differences in evaporation, rainfall, and storms. Enough of the polar ice caps will melt to raise ocean levels and cause flooding in coastal areas including seaboard cities such as New York.

The present global warming seems to shows itself as being real by the shrinking of the ice fields of the arctic and Himalaya mountains and of glaciers anywhere[27]; the glacier erosion is aggravated by the melting water running underneath the ice. Polar bears cannot live without ice floes. The costly storms since the 1970s in the eastern and central parts of the U.S. are more damaging than before by factors of thousands[27]. Their original hurricanes are sensitive to the surface temperature of the oceans because they have traveled over them a long time, through equatorial regions under the sun, from Africa across the Atlantic.

The **ozone** in the earth's atmosphere protects us from solar radiation that can cause skin cancer and eye damage. Nature evolved with that protection such that our eyes have no sensitivity in ultraviolet light (Sec. 1.4). These eyes and our skin too, are now harmed because the protective ozone is diminishing. Over Antarctica, where circular winds confine the region and give chemicals a repeated chance to interact with ozone, one speaks of an **ozone hole**, which is observed between June and November. From year to year, the affected area has been getting larger, as is observed from spacecraft, and its size is already greater than that of Antarctica. The depletion is due to bromines used in fire fighting, and by chlorine compounds used in solvents, refrigerants, aerosol sprays, and foam-blowing agents.

Three types of action are on the way to protect the ozone layer: the **Montreal Protocol** to eliminate production and usage of ozone-depleting compounds, and two rulings by the United States' Congress, against commercial supersonic transports, and against the use of Freon in aerosol cans and refrigerators. (Freon is a trade name for fluorocarbons, which break the ozone down chemically.) Supersonic airplanes produce nitrous oxides as other planes do but at a higher rate and greater altitude such that the ozone is affected more directly; this legislation is bound to come up for revision, because the wealthy want to spend less time in long-distance flights. We note that the military are exempt from any law or environmental consideration; new supersonic planes are being designed.

I have followed the debates over the decades and I became convinced that the problems are real. Even if there were legitimate doubt, we should err on the conservation side because we can afford it now, while the future and well-being of our offspring are at stake. The problems are largely caused by humanity ever since the Industrial Revolution, and we therefore must do what we can to minimize global warming. A U.N. agreement to limit pollutions, the **Kyoto Protocol,** is executed by 140 governments[28]; the U.S. refused participation, but its international airliners may have to buy emission permits. The **worst polluter by far is the U.S.**, followed and surpassed before long by China, which will probably not do anything about it as long as the U.S. does not[28]. The opposition is oriented towards wealth and pleasure, believing that these may be enjoyed, and that the environmental concerns are not real and may be political; I would like to be proven wrong on this observation. We the people are loath to give up our comforts. These feelings are encouraged by the above persistent advertising, which is logical because one has to sell and make profits and promotion. What then is the argument about the future of our offspring? That argument is too remote, as I have experienced when fundraising for protection from asteroid impact (Sec. 12.4). Furthermore, the owners of big automobiles are probably all good parents and doting grandparents, but the salesman said a bigger car is safer.

The disaster - which is sure to come in a few decades - can be prevented only through education of the facts, as is the case for detecting asteroids capable of striking Earth. For **prevention**, it is amazing how much an individual can reduce emission. A home insulation inspection done by an expert can lead to saving thousands of kilograms per year of CO_2 pollution, also by planting trees next to the home and even by making the proper choice of house paints[29]. We must design and build cars with the best possible gas mileage and less CO_2 output (large cars appear to exhaust CO_2 in excess), switch away from CO_2-belching coal plants for electricity generation, increase energy efficiency in homes and offices, etc. All these remedies can be done by making new industries and inventions, while the denial of their effectiveness comes mostly from established industry.

An even more international problem is the burning of tropical forests, which increases the CO_2 drastically, while these trees were good scrubbers of CO_2. The rapid **removal of rainforests** by burning occurs on a large scale especially in Brazil and Indonesia, where major corporations do it. Grassland comes in place of the forest, for cattle, and the beef goes to hamburger sellers, particularly the U.S.

On smaller scales, but occurring nearly everywhere, is the cutting back of the jungles because the local populations have to expand. Madagascar used to have 85% of its surface covered with forest, and now an astronaut in the Shuttle has determined from imaging of the surface that it is only 15%. Clear cutting, of all the trees, is common over enormous areas in the U.S. and Canada, as one can see when flying over. The companies usually plant saplings again, but this has to be kept up by public pressure, and the types of new trees should not be chosen merely for profit.

The loss of forests, swamps, jungles, over-fishing, and U.S. Navy underwater communication at high energies, bring the loss of **endangered species**, the **extinction of species**, which is proceeding at a pace that is frightening because it is an irreversible destruction of bio-diversity. Species interact with other species in a complex ensemble and integration of ecology[30]. It is estimated that some 100 species of plants and animals *per day* are lost forever from cutting the rain forests of Brazil and Indonesia alone[30]. We may be able to sustain these cutbacks for a few more decades, but the end is clearly in sight. The switch to feedlots since the mid-1990s may change this situation by needing less grassland. Thought and care are essential, and protest too, in those countries that are home to the rainforests of course, but even more in those countries where the beef is consumed, where one could decrease the consumption by becoming vegetarians (Sec. 14.2). A special problem referred to above is that the U.S. Navy has begun to use **sonar systems** that are so energetic that they kill whales and dolphins by the thousands in spite of public campaigns and the fact there is not an enemy for submarines to fight in sight[30].

In summary of this section, there are serious global problems of rising temperatures, melting glaciers, shrinking forests, vanishing species, dying coral reefs, and of falling water tables as well as rising sea levels. A source for up-to-date information on the wide range of these topics is the periodical "Worldwatch" and its yearly book "State of the World". The book by planetary scientist Robert Strom appears to be the most professional reference[28]. I can vouch for his impeccable data and conclusions. Strom warns that **global warming is an emergency**[28], and dedicates his book, "To the world's grandchildren. May your parents, grandparents and their policy makers be wise enough to preserve your future". The latter is the key to all global problems, survival of the thoughtful (Sec. 16.3).

I have some **optimism for the long range** because the awareness of the problems is growing, in part through books and teachings like ours. The European Union, lead by its Head, Dr. Angela Merkel, a physicist, decided in 2007 to cut greenhouse gases by one-fifth from 1990 levels by 2020, produce one-fifth of its energy from renewable resources, and set a 10% minimum target for the use of bio-fuels for transport; grassroots efforts are growing in the U.S.[28]. Generally, life has been getting better in the 20th Century, people live longer, they do find solutions and new techniques, and they have become much more aware of the environmental concerns. The driving cause for the problems is overpopulation, and that topic has some good news too, as we shall see next.

13.5 Over-Population and its easy Remedy

Ten thousand years ago an estimated one million people lived on Earth, and by the year 1700 the number had risen to about 600 million, it doubled by 1850, again by 1950, and then again by 1980. Before the 18th Century, the growth of the human population was curtailed by lack of food and medicine, while there were terrible epidemic diseases. But the death rate began to drop with the advance of medical science and education regarding health, hygiene and food production. The latter half of the 20th Century brought the Acquired Immune Deficiency Syndrome (**AIDS**; the word 'syndrome' is used when symptoms are known, but not fully understood). AIDS may have been acquired from monkeys in Africa[31]. The spread of such an otherwise rare tropical disease appears to be due to the increases in modern travel and sexual activity. It can even infect a fetus in the womb. Biotechnology may slow it, but today there still is an increase in

AIDS, especially in Africa, Russia, and India.

The classical question of overpopulation is how long the required increase in the supply of food can keep up with the increase in the number of people. We saw in Sec. 11.2 how Darwin got a start with his evolution theory from the conclusion by Malthus that the living species produce more offspring than can be sustained by the food supply, which causes a competitive selection. We shall however defend in Sec. 14.2 the thesis that people, with their cerebral cortex, can think of new and complex solutions and of inventions in order to provide new resources. This may postpone the Malthus ending at least for some time into the future. For example, **desalinization of seawater** is accomplished through evaporation by solar heating. Its simple goal - but intricate and rather costly execution – is to put it to work in arid regions. The primary problem of overproduction of offspring may be solvable through education in birth control. I have seen that happening in India.

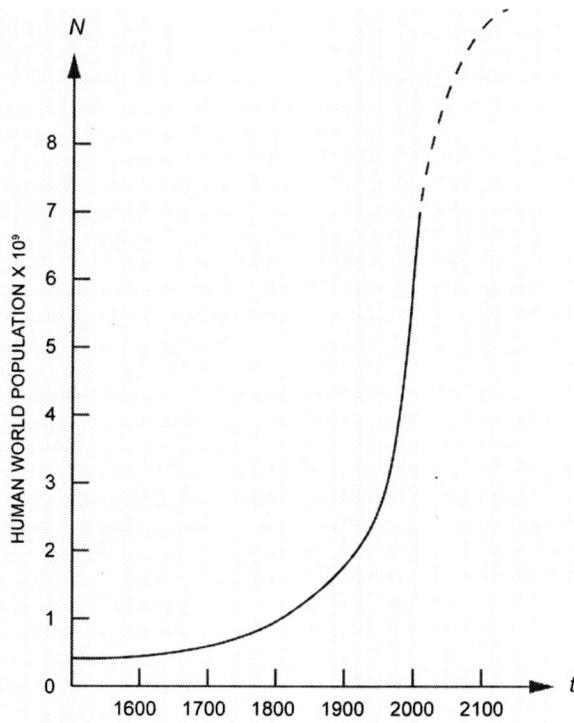

Fig. 13.2. World's population growth. The bend at the top is entirely the author's responsibility, based on the discussion in Sec. 13.5.

Physicist Vikram Sarabhai initiated television for the people of India with a pilot program in the 1960s in his city of Ahmedabad, transmitting to the surrounding villages[32]. Sarabhai's goal was to bring instruction to the people of India in many subjects including **family planning.** Prime Minister Indira Gandhi was active in this cause and Sarabhai consulted also with the Family Planning Foundation. There were some setbacks, it took time for the message to sink in all over India – for the males to realize that protection is not unmanly, and for both parents to realize the benefit of having a smaller family - while Sarabhai realized that it was more effective to transmit to the many villages from spacecraft. For this and other applications, he initiated India's space program which now is a prolific enterprise[32]. And the miracle happened! In the 1990s, the average number of children per family in India had dropped to 2.3 from 6 in the 1960s. The phenomenon is beginning to be seen in other countries as well, with the overall birth rate slowing. That is good news for the future (Fig. 13.2).

The problem of overpopulation is however an intricate one. The medium **population projection** is for 9.1×10^9 people in 2050, a much more drastic downward trend than I sketched in Fig. 13.3, but if women will have, on average, one-half child more or less, the number will be 10.6 or 7.7 billion[33]. The number of poor people is still large and increasing. Furthermore, because there is a general increase in wealth for the wealthier people, especially in the western world, an aging effect occurs in the population such that people have to sustain themselves, or be supported, over a longer life span. To make this even worse, the wealthiest countries have the biggest consumption of resources and they thereby have also the worst polluters. The use of **birth-control** methods therefore remains essential, also in the western world. What surprises me is that in discussions on overpopulation the misconception by Malthus is not discussed and debated[34], nor is the gain in food by a factor of as much as 60 that we will explain in Sec. 14.2. We shall first discuss the related problem of cruelty to people and animals.

14. TECHNIQUES FOR SOLUTIONS

14.1 How to Diminish Cruelty

There is much cruelty in the world that privileged readers are spared from experiencing or even knowing about. The purpose of this section is therefore to select a few of the cruelties present in our world that we might help the less privileged with. First, there are large numbers of school-age children in various countries who are not going to school but have to work in **sweatshops,** and this may be out of dire necessity, because their parents are desperately poor. I have visited a few sweatshops and plantations and seen how adults and children literally sweat 10-hour days in poorly ventilated warehouses. One of the managers had the gall to boast to me that his slaves worked through the weekend in order to make more money, while it was obvious they had to do it to avoid starvation. Clear to see for anyone of us as tourists are older boys and women who break up big stones with sledgehammers for road improvement, and I know that if a stone chip damages the eye, they are on their own. This is **cruelty**. Let us not say that is too difficult a problem to tackle - let us see it as a modern challenge to stop these exploitive and cruel practices.

Information and education can diminish a related cruelty we do not like to talk about it especially when young people are listening. Unimaginable cruelty is the trafficking of women and, increasingly, children. Many are attracted with promises of a good job somewhere, but end up in brothels, in **slavery** without a chance of escape[1]. The number in the United States is near 100,000, about 0.003% of the population; for west-European countries the percentage is near 0.03%, and for India it is near 0.3%. Some of the worst cruelty is in Bangkok to where girls and boys are being sold by their parents from poverty in the countryside, while the western travel industry and American military[2] bring customers to Thailand.

I have had to stop over in Bangkok on my way to India and have seen the open trade at the airport and all over town, "Real cheap, Sir, never been used." Please **think through what your life would have been** if you had been born there and sold by your parents. Use that frightening thought to **tackle** such horrific wrongdoings in the world, to become involved. Let us not go to Bangkok or Amsterdam's red-light district on tourist trips, and spread this message. If our combined actions could save only a few girls or boys, it would at least spare their horror.

Amnesty International is the well-known organization that publishes human-rights violations, organizes successful campaigns of letter writing, and often succeeds in setting political prisoners free. Soviet dissident Cronid Lubarsky, who served five years in prison^3, used to say, "if a few letters came for one of us, he'd be scolded less - tens of letters, he'd be moved to a better cell - hundreds of letters, he'd be free."

Another example of campaigns working successfully is in animal rights where violations are more unbelievably cruel. **The Humane Farming Association** (HFA) accomplished clever activism by placing full-page advertisements in 1989 issues of TIME and Newsweek with a picture of a veal calf in its narrow box. Large letters of the ad brought the reader to attention with a question:

Why can't this veal calf walk?

And the ad answered with:

He has only two feet.

This surely made people stop and read further. So the text explained how the poor beast is kept tightly in a box his whole life, hardly any movement is possible, and how he is *not* fed milk as advertised, but a biotic formula that causes severe diarrhea but that helps make the meat white. That is not all he gets, because further whitening of the meat takes place while the calf is kept alive by reckless use of oxytetracycline, mold-exhibiting chemicals, nitrofurazone, neomycin, penicillin and even carcinogenic, cancer-causing drugs including the synthetic steroid **clenbuterol**. This is not just bad for calves; clenbuteral causes cancer for the veal consumer too.

HFA was sued in court, but has won its case, and veal consumption in the U. S. dropped to one-fifth from 1986 to 2005. The veal case became even more devious and criminal as clenbuteral may not be produced and transported in the U.S. (except for horses with respiratory problems), while there is no such law in the Netherlands. So, for years, employees of a Dutch firm trekked to Wisconsin and other States where veal calves are raised to illegally import the clenbuteral. That was also exposed by HFA and won in court, so one Dutchman was jailed in the U.S. The owner of the company escaped, however, and the U.S. requested the Netherlands to extradite him for serving his jail sentence. Nothing happened, so I wrote a few letters and learned that the Dutch Supreme Court had ruled for extradition, but that it was not executed by a Dutch government administrator, who actually wrote to me that he would not. How is that for Justice? In the mean time in the U.S., the Food and Drug Administration approved a drug similar to clenbuteral for use on hogs, and all producers will be using it, **to stay competitive**[4]. HFA is looking into that and other cruelty, while the veal consumption keeps dropping because more consumers find out what they would be eating. Again, we can help to spread the word. It works.

A documentary book describes the conditions in **slaughterhouses** for cattle, chickens, and pigs[5]. Few people know of the conditions in this industry because reporters are never allowed inside, kept away by armed guards. Everything is aimed at keeping the profits the highest. Cruelty to people again appears as well, because the slaughterhouses generally drive the workers to the highest production rate by extreme speeds of the conveyor belt where they are handling dangerous tools to kill, and sometimes butcher animals that are still alive. Inspectors of the Food and Drug Administration (FDA) are usually on duty, but they seem unable to control the situation or stop sick animals going in. The situation is not any better in the poultry or fish-farming industries - see for yourself the extremely crowded conditions in the ponds, if the farm allows you to enter. You will not get into poultry farms for the crowded conditions are too unbelievable, with beaks cut short so they do not pick each other to death in frustration.

Attempts are made to keep sicknesses of animals in crowded barns below the epidemic level with penicillin – through friends I was invited into one that had a million chickens. The Union of Concerned Scientists has estimated that 25 million pounds of valuable antibiotics, roughly 70% of total US antibiotic production, are fed annually to healthy chickens, cows, and pigs[6]. It improves meat industry profits but puts everyone's health at risk because it is speeding the development of antibiotic-resistant bacteria and may return us to an era when untreatable infectious diseases are commonplace. Increasing **awareness by consumers** will eventually stop at least some of the abuses, and that is where we can help. It apparently worked in 1998 in the European Union when it banned antibiotics that are important in human medicine from use as growth promoters in livestock production[7].

The next secret leaking out into the U.S. consumer awareness is that a profitable and therefore widespread supplement in cattle feedlots is **chicken manure**[8]. Along the same line, Mad Cow disease is occasionally reported, but the media rarely explain the cause, which is reportedly the **feeding of brains** of diseased carcasses. Hear rare testimony from cattle rancher Howard Lyman[9]:

> When a cow is slaughtered, about half of it... is not eaten by humans: the intestines and their contents, the head, hooves, and horns, as well as bones and blood. These are dumped into giant grinders at rendering plants, as are the entire bodies of cows and other farm animals known to be diseased. .... euthanized pets .... the six or seven million dogs and cats that are killed in animal shelters every year..... and roadkill. .... When this gruesome mix is ground and steam-cooked, the lighter, fatty material floating to the top gets refined for use in such products as cosmetics, lubricants, soaps, candles, and waxes. The heavier protein material is dried and pulverized into a brown powder – about a quarter of which consists of fecal material. The powder is used as an additive to almost all pet food as well as to livestock feed. Farmers call it "protein concentrates." ...

Lyman appeared on television, on *The Oprah Winfrey Show,* which was sued on the basis of Texas' Food Disparagement Law (13 other states have such laws too). Oprah moved her show temporarily from Chicago to Amarillo, where it became a local and national sensation - they won their case in court. Some modifications in feed practices and use of animal parts came about in order to help minimize the spread of Mad Cow disease. This shows again that we can decrease cruelty to consumers by speaking up knowingly. Let us now study how we can live better, and protect the environment.

14.2 Healthier and more cheerful Living

We pointed out how human life has been getting better and we will pursue that thought here with the worldwide initiatives and activities towards healthier living, for which there are guidebooks, and I will make some personal observations. The practices of **yoga** and other classical exercises originated in India, Japan and other countries ages ago, but they have now spread through the United States and Western Europe. A promotion of exercise seems to have originated in the U.S. when President Kennedy came jogging from the White House out into the city, never mind his back problems, and it seemed as if the world was cheering and following his example. Even simple improvements can make a difference, such as drinking enough water, but with that one I learned the hard way that a traveler in hot climates must also take extra salts, or the strange sickly phenomena of dehydration may last for weeks (unless cured within hours with a soup spoon full of salt). I have an incidental remark regarding air quality, for I have seen in hospitals cribs with enclosed sides such that the babies are breathing mostly their own output of heavier-than-air CO_2.

The U.S. makes a half-hearted attack on **smoking** in the U.S. since the 1980s, also by the airlines for having to do less filtering of the air in their planes. Smoking is stimulated with taxpayers' support for tobacco growers. These monetary subsidies happen even though there are non-smoking rulings in public buildings and restaurants, together with required printing of cancer warnings on cigarette packages. No warning is required yet on the packages for the **slavery aspect** of smoking, that it is the result of an addiction to the numerous ingredients put into cigarettes. The tobacco industry intentionally uses chemical additives, which facilitate and maintain addiction, ameliorate the harshness of tobacco smoke, increase the effectiveness of nicotine, and the industry also allows unintentional additives from pesticides, agricultural chemicals, soil bacterial flora, and manufacturing processes to enter into the final product[10]. The information is brought to our awareness in a movie called "The Insider", Dr. Jeffrey Wigand, who is one of the few courageous enough to come forward and suffer for exposing a giant industry. But on a University Campus it is deeply sad to see young people - aspiring leaders of society - hooked on cigarettes, enslaved by a deviously moneymaking complex. The addiction is sustained all over the world to an alarming extent, with even children smoking everywhere (except in planes), and the warnings for cancer are not even printed on the packages abroad. A World Health Assembly had 192 nations sign up in 2003 to discourage the cigarette smoking from which an estimated 5 million people die per year. The U.S. and China, both large tobacco producers, did not sign.

As for **exercise**, each person should find her or his own regime, and **massage** may be a part of it. I for one do not exercise at waking up, but do some **pervasively** during the day. This includes stops for a rinse of the eyes, massage of eyes, ears and cerebral cortex, and having fun breaks for shoulder stands, which bring refreshing sensations because of the change in blood circulation with exercises for the legs and massages by the legs to

212   TECHNIQUES FOR SOLUTIONS

each other, doing that in the shower as well[11]. (See your doctor first. Head and shoulder stands are safe only if your spine is up to it – be careful!)

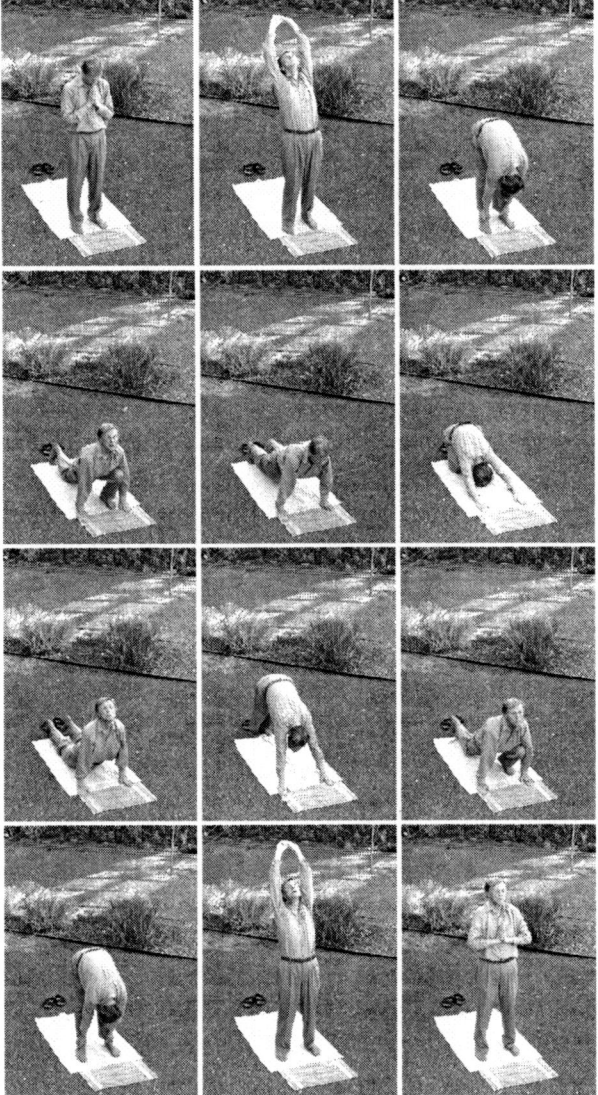

Fig. 14.1. Tom Gehrels doing the 12
stages of the Sun Greetings.

A set of eight or so of the 12 stages in Fig.14.1 per day keeps one lithe for life; breathe out at the end of each step. These are basic and should be done regularly, and then there are so many opportunities for exercising, walking, swimming, and biking, done simply but pervasively.

The industry that hangs glasses on people does not seem to endorse **eye exercises** – I found it hard to find a teacher. There are cases where exercise can take care of a problem - the challenge is to find the rare eye specialist to approve and give advice for these exercises since they must make sure there is not a medical condition present that could be aggravated by the activity. My favorite argument against glasses is that if the muscles of a leg are getting stiff, one does not put that leg in a cast, and so it seems for the muscles of an eye. I have described various eye exercises elsewhere[11]. It is said that schoolchildren and certain industrial workers in the Peoples Republic of China have pauses for eye exercise and relaxation.

Stress is the cause of many ailments, while **depression** is the professional hazard for scientists, managers and others who have extreme excitements or stress-creating responsibilities in their daily work; Darwin may have been the classical example (Sec. 11.2). Heart specialist Ramesh Kapadia of Ahmedabad, India, has become well known for doing **stress-reduction** exercises with his patients and for publishing guidebooks in this discipline[12]. Some 200 of his patients, including those who have had serious surgery, do an hour of training twice a week with a yoga instructor.

Meditation must be millennia old, and easy for anyone to do. Kapadia and his instructor fold it into their hour of training and literature. Try it if you like, have a session, lying flat on the floor. Concentrate on the breathing, using the midriff rather than the chest and **feel the air** going in and out deeply. This already is an exercise in holistic healthcare since breathing supplies the life-giving molecules within us. Train yourself to keep doing the special breathing during the meditation automatically. At this stage, you must also train yourself to eliminate any other thought during the meditation. Now begin the concentration on successive parts of the body; you may move each of them a little as you come to it. Start at the tip of your toes. Come up with your concentration through feet and legs. Travel slowly and consciously to inside components. Eventually move up to the chest, heart, and lungs. Travel out with your concentration into the arms and fingers, one-by-one, slowly. Relax. Move the concentration through the neck and the back of cerebrum and cerebral cortex towards the top of your face. Relax. Slowly now, force your extreme concentration and imagination OUT through the spot between the eyes where Indian women have the red dot called Bindhi, or Tilak. Now try to move your concentration slowly upwards, and look back on your body lying there, totally relaxed. Go anywhere you like - I go to an asteroid or comet sometimes. Eventually let yourself come back and awake. The practice of meditation brings enhancements of peace of mind, greater will power and concentration. I do not adhere, however, to anything supernatural; we do the exercise and meditation routinely and improve them gradually as an enjoyment for body and mind. However, there is a **holism** about such healthy and cheerful living, whereby nature is seen as a whole that is more than the mere sum of the interacting parts, which is a characteristic of evolution (Sec. 7.1).

Such a holistic lifestyle includes being a **vegetarian** and it connects with the previous Chapter because there is an urgent need in the world for vegetarianism as it saves resources to a factor of about 60. The number is approximate but logically well established, namely from eating grains directly, rather than processing them through animals, and being more efficient by a factor of 60.

Discussion of vegetarianism in class is an odd topic because there are many reasons for it and yet **people seem reluctant** to even consider it. I can explain this only as due to people growing up and developing eating habits that are molded through advertising to an overpowering extent by the food industry, particularly to protect their hugely profitable product, which is meat including fish, veal, caviar, chicken, and pâté de foie gras. There are hardly any reasons to be against vegetarianism, other than that vegetarians must take care of a shortage of proteins by supplementing with tofu.

On the other hand, the advantages of vegetarianism seem to be overwhelming. The above advantage of the factor of 60 is dominant, but vegetarians also diminish the demise of rainforests and the suffering in slaughterhouses and overcrowded cages, fishponds and feedlots. They avoid the squandering of wonder drugs to fatten livestock and prevent their epidemics[13], and the daily injecting of milk cows with Bovine Growth Hormone (BGH), which is done for a 15% increase in profit[14]. The cows given BGH have a high incidence of udder inflammation, and antibiotics are used to treat that. It is pathetic to have notes posted right by the counter, not to eat more than 6 ounces of tuna per week because of mercury **poisoning**, and similar ones for most fish[15]. And so it goes on to make more profit in violation of consumers' trust and health. A **vegan** is a person who does not use anything that comes from an animal, no eggs, no cheese, and no leather. An advantage of vegetarianism is the liberation from **slavery** to the meat industry, its cruelty and its advertising. The counterpoint regarding profits is that they enhance a society, which is however already wealthy enough to be eating meat.

More and more alternates become available at least in the U.S. such as soymilk, eggs from cage-free chickens ("scratch chickens"), and there are environmental farms and ranches. Vegetarians and vegans feel lighter, are healthier and hardly ever constipated, they save money because vegetables are much cheaper than meats, they are less likely to suffer cancer, or heart attack, need not be afraid of Mad-Cow disease, and are proud that they actually **do** something towards improving the world. They like to experiment and are used to variation in their meals for reasons of taste, but also to give the body the best chance to obtain what it needs; the great advantage of that lifestyle is that medicine needs rarely be prescribed. The first thing to do in fighting **obesity,** which is a growing problem in the U.S. and elsewhere, is to become a vegetarian. Anyway, we shall go on to other modern improvements.

14.3 People find Solutions and make Resources

Our relationships with the environment and other people are becoming more interwoven and united in an **ecosystem,** which is a community where the relationship between organisms and their environment is a unit (Sec. 13.4). For example, forests play an essential role by cleaning the air, sustaining a variety of species, and moderating local, regional, and even global climates. People have also become a part of the ecosystem. We mentioned how Malthus held forth that the increase in human population would run into a shortage of food and resources. But people *make* resources as well as consume them, and the cerebral cortex allows them to find solutions. The profit motive also stimulates finding new solutions by driving the invention and production of new foods and facilities. An example of people finding complex solutions is the asteroid problem and what we are doing to mitigate it in international collaboration (Sec. 12.4).

The following is an even more sophisticated example of people making resources, which will make a big difference for the future if it succeeds. There are expensive projects underway in various countries to develop powerful **fusion** generators, which would not leave radioactive waste behind, as fission does. Fusion energy is therefore of great interest, but it is extremely difficult to reproduce the required conditions of pressure and temperature as high as in the interior of stars. There are experiments with compressing the fuel with laser beams, or by magnetic effects. We may expect results from the studies and experiments in coming decades; people often have doubts of progress in the beginning of a big project, but then it suddenly bursts forth in the development of sophisticated technology.

The mining of resources on **ocean bottoms,** at intermediate latitudes, is another example for the future as there lies a wealth of possibly organically grown spheres the size of apples consisting of precious cobalt and manganese. Who owns or has the rights to this resource? Who will have the financial capability to mine it? The U. S. has refused to sign the International Seabed Treaty of 1982[16]. Even more remote and costly examples are the very early studies of zero-point energy (Sec. 14.6), and the exploration of asteroids in space (Sec. 14.7).

Recycling of valuable materials among our wastes is increasing worldwide, but many restaurant or business owners in the U.S. seem not to want to be bothered yet because it is not economically worthwhile to them. Concerned citizens can be quite effective in urging them to join us and their city government, and in encouraging everyone else towards participation.

Personal action is also essential in the **conservation** of energy and promoting others to do the same[17]. So many ways are possible, ranging from turning off lights when not needed, to choosing energy efficient cars and appliances; hybrid cars are becoming more affordable and it is fun to drive them and to see the savings displayed. It can become a daily habit of sizable impact if large numbers of people participate in conservation, and this is increasing. Periodicals provide guidance on recycling and renewable energy such as **wind energy,** literally with windmills that are coming back into usage in modern form[18]. Denmark works on the goal to generate 50% of their electricity needs by the year 2030.

An alternate source of power is solar energy, which is obtained with photovoltaic solar cells, which convert photons into electrons, light into electricity. They are crystals of silicon or gallium arsenide that transform radiation into a voltage difference. Many crystals are put together in large arrays forming a **solar panel** for spacecraft in the inner solar system, not too far from the sun. They are generally used and apparent also in places such as on the roofs of private homes and on the fields of solar farms. Our pride is the set on our roof, which appears to being paying off in about 8 years as the electric company buys our excess electricity. With increasing demand and supply comes lower cost and efficiency such that solar energy is competitive with electricity from fossil fuel plants, and from hydroelectric plants, which use the drop in water levels made by dams in the rivers. The price of solar cells and panels is coming down (Sec. 14.4).

The great advantage of the electric motor is that it **inverts its role** of electricity user into that of a generator using the braking energy into a charging gain, the perfect case of reciprocity. When solar panels generate an excess of electricity for their application such as a home, it is put into the commercial network; electric trains do that too, their braking is converted into electricity and saved, while **electric or hybrid cars** do that too. The former relies entirely on an electrically charged battery and an electric motor, and the hybrids have partial reliance on a gasoline-driven charger.

Fuel cells hold a catalyst to perform a chemical process without moving parts that combines oxygen from the air with hydrogen fuel to produce electricity and water, the car engine is electrical. Hydrogen is abundantly available in nature and can be made from a variety of sources such as biomass (residues from forestry or agriculture); it would be ironic however if oil and natural gas were used in the process, as some propose. As a fuel, the biomass oxidizes, "burns", to water, and the energy can be made available to drive cars. The water vapor comes out from the exhaust, rather than the usual CO_2 or other poisonous gases, to diminish pollution and greenhouse gases. These techniques have been tried in military and space applications and now, after the exploration of resources other than oil seemed to have been generally resisted in the Congress[18,19] for decades, **hydrogen-fueled cars** are slowly being developed in the U.S.[20].

In the U.S., there rose in 2005 a sudden development to use **ethanol**, C_2H_5OH, instead of gasoline in automotive engines designed for such usage. It is the intoxicating agent in liquors and it is a solvent, so that many people know about it; it is also called **ethyl alcohol** or **grain alcohol**. It is now promoted to make the U.S. independent of foreign oil as it is readily produced from corn kernels, and there may be more efficient ways to produce it from other plants or parts of plants[20].

Another new development is the use of **robotic devices**. In the military, this can of course save lives and the switch to fully computerized fighting aircraft, without people on board, is therefore amply seen in aviation literature. The usage of robots is penetrating everywhere such as in farming, industry, and even in the household[20]. The remainder of this chapter will bring even deeper sophistication in finding solutions.

14.4 The Human Genome and Genetic Engineering

A modern example of people discovering resources in full swing is in genetic engineering. In this section, we continue from the discussion of the nomenclature in Sec. 11.3, where we saw the basic structures of **DNA** consisting of long molecular chains in all living matter. Proteins package the DNA into chromosomes and the cell does further packaging but with **splitting and recombination** to achieve reproduction. Our binary world is in full swing (Sec. 15.6).

There are many groups of scientists with their projects all over the world to read the DNA structure of a whole organism, to read its **genome**, the set of genes which are stretches of DNA containing protein instructions, passed down from cell to cell for each organism[21]. The whole set of instructions for a person, contained in their 23 pairs of chromosomes, is the **human genome**. The genome projects identify all these letters and groupings with the instructions they brought to the body.

The **Human Genome Project** is an international organization with the goal to determine all the human genes; supported by a variety of organizations and pharmaceutical firms who are interested because there is potential for development of new drugs. In the race to complete the description of all human genes, commercial projects matched or even surpassed the Project, but eventually an overall agreement and arrangement was made, and that was a momentous occasion[22]. The importance is to know the place where the instruction for a defect or sickness is in the chromosomes; illnesses can be identified in terms of mistakes in the sequencing of the DNA, and perhaps be corrected, or the disease may be tackled more knowledgeably[23]. A revolution has come upon us in fighting diseases and perhaps in breeding selectively and thereby directing evolution. This is why the 21st Century may become the one of biotechnology, compared to the 20th Century as the one of physics.

Genetic engineering is interfering with the DNA instructions; it is the directed manipulation and application of genes by making changes to their instructions. It has become possible to identify a single gene with a single instruction in an organism, and to study, alter, and transfer it to another organism. When one extracts the DNA from a tissue, strands of its thread are seen; the structure of the DNA is resolved by ultra-high resolution microscopy and crystallography. An often-used method for putting the genes into another organism is by micro-injection, whereby small fragments of the desired DNA are literally shot with a gene gun through the plant or animal cell walls into the nucleus of the receiving cell.

A variety of genes can be added to **seeds**, for example for protecting against the effects of pests, frosts, or excessive solar radiation. Genetic engineering is to produce something we want, such as insulin, a better beer, or a protein for food. To do this, one must change the set of codes in the DNA; this is regularly done with plants and animals, for instance by changing genes or experimenting with new genes in mice.

A beginning is made for people in some experiments to treat diseases, for example by injecting viruses that contain the gene that needs to be changed. The virus will infect host cells and change the DNA code; in nature, the HIV virus uses a similar technique to infect humans. Regular diagnosis is being made for prevention of sickness and death by early treatment based on study of the genes, which now can be done for anyone. For example, the likelihood of getting **breast cancer** or **prostate cancer** is indicated by a deletion of a C at nucleotide 7471 in Exon 15 of the BRCA2 gene (c7471delC; BIC nomenclature 7699delC).

The major commercial application is towards the improvement of plants and fruits, making them resistant to the pests that would otherwise kill them. One can also make plants acquire nitrogen by way of genetic engineering instead of needing fertilizers, as they already do naturally for CO_2 from the air. This sounds great for making profits in farming, but consumers may not be convinced that it is safe or healthy; or, they may simply not be told what is being done. Much of this is happening without our consent or even knowledge, it happens for motives of profit. American manufacturers are not required to reveal whether or not their products have been genetically modified, so we do not know what we are consuming. The benefits of genetic engineering are many, such as reducing the use of pesticides, or improving the food value, and the product usually looks better. We must also take into account, of course, that genetic **modification** has been going on for a long time whereby better products are favored, and the others no longer produced.

It may be dangerous to play with the ecological balance of nature[23,24]. The reduction of pesticides sounds good, but nature has a way of reacting so that new types of weeds may develop along with the improved crops. By about 1997, perfect-looking fruit appeared on the shelves of stores in the U.S; by the year 2001, 60% of the processed foods in America's supermarkets contained at least some genetically engineered ingredients. We have been eating the products for many years, and no reports of ill effects have come in, as far as I know. There have been protests abroad against their import of genetically altered food from the U.S., and protest within the U.S. have demanded safety testing and clear labeling by the American Food and Drug Administration[23]. Another danger is the misuse of **gene doping,** by inserting genes for permanent muscle enhancement, which menaces to change the nature of sports[23].

Human genetic engineering is a process to change the code of a gene for a person. The treatment of the patient for a defect is done without affecting his sperm or her eggs, it is **somatic modification**, generally considered acceptable; the word somatic (Gk. somat, body) means that normal cells, not "sex cells", are involved, and nearly all our cells are somatic; germ cells is the usual name for sex cells, the word derives from germination. But the genetic modification can also be done in "inheritable genetic modification" or **germ-line engineering** so that is with non-somatic or germ cells; this procedure changes the genetic composition of the child to one that differs more than normally from her or his parents, it changes its nature and that of its offspring. This has been done for animals but not for people as yet; it has been proven on animals that DNA instructions can be taken from the one and injected into the other, but not yet for people (as of 2007). The genetic duplicate of a species is called a **clone** (Gr. klon = twig), where DNA material is transported into a fertilized egg; after such cloning, the embryo is implanted in the uterus. This process has been banned for people in over 30 countries. A few claims of births of cloned babies have been made, but also been doubted because the procedures are intricate and the chance of failure or partial failure is high[24]. It is now accepted that this procedure never succeeded, and that the claim was a fraud.

The name of **stem cells** comes from the fact that all other cells stem from this type of master cells that have the potential to develop into almost any type of tissue[25]. They are found when a fertilized egg begins to divide producing the embryo; stem cells are also found in bodies of infants, children and adults but these can then only make the tissues they came from. Therapy using embryonic stem cells offers great potential for repairing diseased or damaged tissue (as in Parkinson's disease). All of this brings exciting new biotech industry, but one can see that there is danger of misuse or of usage that may go wrong[25]. Here comes another case of pros and cons.

14.5 Molecular Nanotechnology

Exciting new industry combined with danger of misuse may also occur in molecular **nanotechnology.** It is technology at sizes near 10^{-9} meters, a billionth of a meter; so that he word "nano", 10^{-9}, allows us to understand the word "nanotechnology" at a glance (the word is in the Table of Note 2 for Chapter 1). The complex molecules one deals with in this technology have as many as 10 or 20 atoms, so they are about ten times larger than atoms, which have sizes of about 10^{-8} centimeters (10^{-10} meters), and their own size is therefore near 10^{-9} m, a "nanometer," 1 nm.

The nanotechnology process makes these molecules reproduce themselves, or fabricate others like them[26]. Reactive molecules are catalysts, stimulators, and they are positioned in mechanically controlled places such that the reactions and complex structures can be controlled. This technique is already **successful** in making better components for circuits in electronic industry. If it can be proven safe, there is no end in sight even for hitherto undreamed solutions in medicine and engineering. For example, printable solar cells may become available at low cost and that would provide cheap energy in many places, for small applications and for large fields of solar panels (Sec. 14.2)[26].

However, there is the serious question regarding the safety of the techniques, how well the reactions and structures can be controlled. If the controls would fail, there could be health risks of new particles with unknown properties, such that suggestions ensue for **banning** this type of research. Could one really succeed in banning any intriguing and lucrative research? That question seems answered at the end of the previous section: one may declare that the cloning of people be banned, but one doubts that the ban is successful everywhere. Somewhere, sometime, a violation will come. The situation with molecular nanotechnology being lucrative is similar to that of nuclear engineering, which is used to great advantage in many applications even though it is dangerous. In this case, strong financial support for the new technology is easily forthcoming because of the possible financial success of medical applications[26].

Nanotechnology in the hands of irresponsible people does seem to be a dangerous pursuit, and there is always a chance of accident in a genuine lab when working with replicating assemblers. It is the danger of an unstoppable self-replication. Tiny, destructive **nanorobots** are fearsomely foreseen, that would replicate by feeding on the environment – here I am quoting from literature, but have a problem understanding just what could be edible, digestive or useful in construction. It is feared that they could spread swiftly and be too small, tough and rapidly reproducing to be stopped.

However, the engineering is intricate and complicated. The situation may be similar to that of the Atom Bomb project at Las Alamos during World War II, when the theoretical principles were known and rather simple, but the execution was **much more difficult** than had been foreseen. In that example also, it was feared that the reactions would runaway unstoppably. Nanorobots will not be easy to build and it seems unlikely that they would come about by accident and then be uncontrollable. However, warnings are important, for it may be risky engineering, and awareness can help reduce or eliminate the chances for abuse or failure.

Sir Martin Rees, Astronomer Royal of Great Britain, took the threat seriously. He wrote a book with a host of information for taking care of our future[27]. It is, however, a pessimistic volume, *"Our Final Hour"*, in which he gives odds not better than 50-50 for our society to survive the 21st Century. It reminds me of *"Essays on Population"*, written by Malthus, in the same college as Rees', holding forth that the increase in human population would run into disaster for lack of food. Malthus' and Rees' dire predictions seem wrong by not taking into account that people find solutions and make resources. Rees proposes that severe restrictions imposed on nano-research. Charles, the Prince of Wales, shares his concerns, and speaks about that, which brings an interesting political controversy because the British Government supports nanotechnology financially, in the belief that cures for diseases such as malaria and tuberculosis might be possible. It has become a jolly debate. In contrast, we will next see a dreadful case of controversial history.

14.6 Space Exploration, after gruesome pioneering

Here we switch to the exploration of space, by actually having instruments and people **in** space, rather than at astronomical telescopes on Earth. Powerful rockets have to overcome the gravity of the earth; the escape velocity is 40,000 km/hr, more than a hundred times faster than the fastest trains (see the discussion of escape velocity in Sec. 5.4). The first were **sounding rockets,** for sub-orbital flight, produced in 1944-45 for German warfare. Its haunting fact is that the workers were beaten and starving slaves, many of whom the finest individuals who had resisted the Germans in German-occupied France and other countries, and had been caught. They hated the Germans, and even in the factory and its concentration camp, they were sabotaging the work as much as they dared, knowing that they would be tortured and hanged if caught; the 'hanging' actually was a **choking to death**. Figure 14.2 was drawn by one of the political prisoners after the war. Before the hanging, their torturers wired little sticks into the rear of their mouths to prevent them from shouting a last defiance. I have seen the sticks in the museum, with tooth marks, but they were gone on my next visit, stolen by neo-Nazis who had broken into the museum[28]. Dora was the name of this camp, next to the underground tunnels in the Hartz Mountains of central Germany. Most of the German camp guards had been criminals released from prison for this purpose, and they had learned techniques of ruthlessness in a training camp of their own. Now they had guard dogs and traitor prisoners; they were never to show empathy with their prisoners, not even after being together for months[28]. The museum and the underground factory tunnels are open to the public near Nordhausen, about two hours driving northeast of Frankfurt Airport.

It behooves us to recognize and **honor the pioneers** of whom many perished - 20,000 of 60,000 prisoners died at Dora - as we benefit from worldwide space programs. The irony of the human predicament is that something so profoundly important for its future should have emerged from such unbelievable cruelty to fellow man, and that our colleagues presided over it. Wernher von Braun was the leading engineer, and his brother also worked at Dora. Born into an aristocratic anti-Nazi family, he was fanatically dedicated to the dream of space flight. Eventually, he was to become an honored pioneer in the U.S.[28], who designed the largest of rockets

Fig. 14.2. When prisoners in the concentration camp were caught sabotaging, they paid the price and all other inmates were forced to watch.

to begin the flights to the moon in 1969, and his 1977 funeral service was in the National Cathedral, Washington, D.C.

The change in von Braun's allegiance occurred immediately after World War II, when the remaining Dora rockets were rushed to the U.S., where they were launched and flown. The von Brauns and 125 of their wartime associates had also been rushed to the U.S., to speed the race into space with the Soviet Union, which they did[28]. For years, the Soviets were ahead, with **Sputnik** (traveling companion) as the first Earth-orbiting spacecraft[29] took flight on October 4, 1957, a Laika-breed dog following within a month, exploring Venus and Mars, and by entering the first man and the first woman in space. However, the **race for the moon** was on, as we shall describe in the next section[29].

Space flight was to bring the greatest advance in the study of planets, stars, our universe, and their origins. An increasing flow of information from the space missions and from the study of meteorites brought advances such that one can now study our origins in detail and rigorously. Rocketry became an **international enterprise** for communication, precise global positioning, relaying of data between satellites and earth stations, reconnaissance for military or weather purposes, and for spaceflight of people[30]. There is also a lot of debris in space, from so many launch rockets, staying up long times, but sometimes crashing down into the earth's atmosphere.

Space has an abundance of **solar energy**, greater than on Earth because of the availability 24/7, meaning all the time, and the absence of shielding of short and long wavelength sunlight by the earth's atmosphere. Spacecraft use panels of solar cells as we mentioned, but for distant exploration of the outer solar system, **nuclear energy sources** are used, in spite of concern about launching - and then perhaps crashing - the radioactive devices.

There are potential resources in space that have not even begun to be explored for possible utilization, like those we may find through **asteroid exploration**[31]. Studies for missions to comets and asteroids have been made since the early 1970s[31], a few automated missions have already been executed or are underway for making observations and analysis of surface samples; the samples can also be sent to Earth and down by parachute for more detailed study. As for future mining, consider a metallic asteroid, similar in composition to iron meteorites with high concentrations of iron, nickel and the platinum group of metals (platinum, cobalt, iridium, osmium, and palladium). On Earth, these heavy metals are not available because they differentiated, sinking out of reach towards the center during the early stages of formation. The same differentiation occurred for large asteroids, not during their formation but later through radioactive heating. Then they cooled and solidified in differentiated structure, and later some of them were broken up in collision such that the metallic core fragments became metallic asteroids and meteorites. Some asteroids show evidence of crystalline water so that one considers mining in support of space colonies[31].

An expensive exploratory program would, however, be needed to develop the techniques of mining and processing in such remote and low-gravity environments. The basic question emerges whether or not **private industry** has the resources or is willing to take the risk of large investments into long-range future possibilities. Because of the long preparation for such ventures, industrial companies are reluctant to get involved. The U.S. Government therefore did it, with as many as seven NASA centers created for such studies and planning, but only for a while, until the funding stopped due to a reversal of congressional opinion on the practicality of such endeavors. Military interests can lead us into space, but only for specialized goals such as reconnaissance and perhaps weapons in space. Public interest and involvement with international cooperation may be the answer for limited enterprises[32]. People in space are next.

14.7 Human Space Travel

The first person in space was Soviet cosmonaut **Yuri Gagarin**. The first footstep on the moon was by American astronaut **Neil Armstrong**. That was the result of an openly announced challenge by the young American President John F. Kennedy to the U.S. Congress in 1961 and it sounds like poetry when you read it out loud:

> "I believe that this nation
> should commit itself
> to achieving the goal
> before this decade is out,
> of landing a man on the moon
> and returning him safely to Earth.
> No single space project in this period
> will be more impressive to mankind,
> or more important to the long-range exploration of space,
> and none will be so difficult or expensive to accomplish."

His proposal received more doubt than enthusiasm, but after the murder of this President shocked the Nation, his successor, President Johnson, had no difficulty putting Kennedy's dreams into quick execution. How close the **race to the moon** had been we learned only after the end of the Cold War, how the Soviets had had a major accident and setback of a huge rocket blowing up on its launch pad. One might therefore wonder if the U.S. had been first on the moon if President Kennedy had not been shot.

The **Apollo** program conducted the moon landings with the largest of rockets, produced under the guidance of Wernher von Braun[28]. The rockets had an array, of jetting stages to bring a lunar module for landing and sojourning on the surface. Neil Armstrong's piloting and stepping onto the surface – cool as a cucumber, never skipping a heartbeat - had indeed occurred before the decade was out, in 1969. The United States proceeded to build the Shuttle vehicle for flights of several days and then return and land like a plane. Russia has similar ones, but landing by parachute in the deserts of Kazakhstan; their pilots and crew are **cosmonauts**. The People's Republic of China has entered the space program vigorously[33]; its people trained for spaceflight are **taikonauts**.

For long-duration flights, there are **space stations**. The Soviet Union, now Russia, had MIR ('peace') that was regularly visited by cosmonauts and astronauts. After 15 years of service, the decommissioning of MIR occurred in 2001 by crashing it into the earth's atmosphere and ocean. The International Space Station (ISS) took over as a truly international enterprise although the U.S. is withdrawing in favor of its moon and Mars program, which brings uncertainty regarding the future of that enterprise. The ferrying of people and equipment done for a long time by the Soviet, now Russian, Soyuz, will be replaced with a clipper type vehicle in a joint development with the European Space Agency. The U.S. is replacing its Shuttle fleet with vehicles that can fly to the moon. Typical visits to the ISS are for a few months, but can be as long as a year. That is for quite a few sunrises and sunsets in 90-minute orbits! These are all for "low-Earth orbit," staying safely underneath the protection from energetic solar radiations provided by the Van Allen belts. Russia has even flown tourists to the ISS, at a price of some twenty million dollars each; the U.S. has had a few political guests in the Shuttle. Tourism and burial in space are soon to come. There is an interesting race to the moon brewing again, with China expected to land their taikonauts. The U.S. intends to do so too, by the year 2020, perhaps in preparation for going on to Mars by 2027. India may also land people on the moon.

From a scientific point of view, it is much simpler and safer to fly **instruments only**. There is a continuing discussion for or against space flight by people, especially after two disastrous Shuttle accidents. But there will always be space flights by women and men because of their **desire to fly**; it seems an inborn hankering of people. So we have both types of space flight, by people and by robots.

The **combination of the two techniques** put the Hubble Space Telescope (HST) into orbit by the Shuttle, which occasionally returns to the HST for a maintenance mission, done by astronauts in spacewalks. It has been possible with available Shuttle techniques to deliver an automated spacecraft to Earth orbit and this robotic craft then proceeds to Jupiter and, with a gravitation flyby assist by Jupiter to Saturn, or land on asteroid Eros and on Comet Borrelly. It is not necessary to have a manned landing on a metallic asteroid - the mining would be more economical with automatic equipment than by astronauts.

Mars is an attractive target in connection with the study of the early stages of the solar system and of the possibility for finding primitive life forms; there is likely to be water and perhaps organic matter below its surface, although the dust storms might generate enough electrical charges to affect the molecules[34]. There is great activity in space exploration of Mars and this includes at least consideration of people going there, even though their pollution might affect the search for primitive life forms.

However, a round-trip to Mars may take as long as fourteen months. The greatest danger for human travel beyond the protective Van Allen belts is the **energetic particles** or radiation we discussed in Sec. 1.2. From tests made on mice, the conclusion was that the effects are serious, that sterility might ensue. The Apollo flights to the moon in the 1960s encountered no solar storm; in any case, the astronauts were not supposed to produce more offspring. Cosmic rays would slice one-third of the DNA for each year spent in interplanetary space[35]. Engineering studies show that one must have a protective layer around the people's cabin of perhaps as much as 4×10^5 kg, depending on the material, but putting great demands on launch capabilities[35]. The moon or a near-Earth asteroid might provide such shielding, their gravity is lower in lift-off, and one might obtain water from crystalline components of the soil, but all such mining and separating is costly. There are other places in the solar system where people could land, such as on the satellites of Mars, and on asteroids anywhere. The four large satellites of Jupiter are, however, inaccessible to people because of the deadly radiations near the powerful Van Allen belts of Jupiter.

An appreciable amount of energy is essential for space flight of course to provide heating for people and power for electronic equipment. One can **recycle** to reduce the amount of required water and oxygen; from the CO_2 exhaled by the astronauts it is possible to recover oxygen, much of the needed water can be recovered from wastes and from evaporation, and it seems feasible to grow food from waste products. We have already discussed solar panels and nuclear energy generators. Scientists and engineers with a feel for the future have proposed **ion propulsion** for decades, and the first spacecraft are now flying with it. The molecules of a gas, usually xenon, obtain an electrical charge by removal of electrons, when they are ionized[36]; the ions have the same positive charge, they therefore repulse each other, and the resulting gas comes streaming out as from a small jet engine. For example, the "Deep Space I" craft used it to fly by Comet Borrelly in 2001, and Japan's Hayabusa explorer used ion propulsion to observe asteroid Itokawa, which yielded the first observation of a rubble-type object without impact craters[36].

For a yet farther ranging and more powerful propulsion of spacecraft, and possibly for energy generation on Earth in general, studies are underway of **Zero-Point Energy**, which is also referred to as electromagnetic quantum-vacuum energy[37]. With the zero point is meant absolute zero Kelvin, 0 K, which we believed until recently to be the end of all activity, a temperature that could not even be reached, but it was actually predicted in quantum theory.

For the possibilities of extended existence of people in space, experiments have been made on MIR already, with nearly a year of living without gravitation, in a **zero-G** environment, which is also used for experiments of processing metals and growing crystals in space. On the nuisance side, zero-G causes nausea; nearly all astronauts suffer from it during the flights but they do not like to talk about it. Loss of balance is severe for some people, with difficulty in standing and walking for a while after return to Earth. The worst may be a puzzling problem of calcium loss in the bones from long space travel.

I have a curious consideration of manned flight. The flights to and the walks on the moon were shown on international television, in newspapers and magazines, watched and discussed by everyone on Earth. The astonishing feat of people leaving Earth may have initiated the surge of activity and questioning in nearly all countries at about the same time of the 1960s, The Great Sixties. If we could go to the moon, there seemed to be no limit: **"If we can accomplish that, we can do anything!"** I believe it was reminiscent of 1783 (the year is easily remembered by 17 and -17) when the first-ever **balloon flights** were made, and they were executed from the squares of Paris, four flights during that single year, with the population watching in excited crowds. People were free to leave the earth! What other freedoms were within reach?! And the French Revolution began soon after.

15. ADVANCED LIFE ELSEWHERE

15.1 Searches for Planets of Other Stars

The topics of this chapter are the greatest, treating evolution in the broadest sense. We will need all that we discussed before. Our imagination goes first out to stars we expect to have some sort of planets[1], we saw such planets in the formation process of the sun in Chapter 10. Why would we not see them when interstellar clouds condense into stars? The main observational problem is that the planets are too faint for direct observation, because they radiate only **reflected light** from their star, and when seen from a distance they would appear too close within the glare of their brightly shining star. However, such resolved imaging is attempted with special techniques. Detection of planets orbiting other stars therefore became a special challenge to observational astronomers that they began to take on in the second half of the previous century.

There are classical indirect techniques, whereby one observes the **star,** instead of the planet, for signatures of the motion of its planet. We understand for each planetary system how the gravitational bond makes the planet move in an orbit about the star, but the key realization is that, in reciprocity, **the star will also move** around a little because the planet's gravity pulls it also, just a little. Both the planet and the star describe their own orbits about their common center of gravity. The star will however move very little because it is so much more massive than the planet[1].

We will consider two different geometries of observing the system. First, consider a system that has the plane of its planetary orbits (the ecliptic plane) **perpendicular** to our line of sight. A massive planet (such as Jupiter) will then pull the star to move in a small circle projected on the background of the sky, and that period of revolution of the star will always be the same as the orbital period of the planet. That is 12 years in the case of Jupiter and, in this geometry, the small motion of the star shows at various times in an ellipse or close to a circle over 12 years.

Now imagine the opposite orientation, a planetary system with its ecliptic plane **along** the line of our sight; we then see the system projected edge-on, as a straight line on the plane of the sky. Now we cannot see the star describing an ellipse, but only that straight line. However, in addition there may be observable motions **towards and away from us**. Perhaps even an **eclipse** of some of the light from the star might occur when the obscuring planet comes between star and observer. Such transits of other stars, of a cluster perhaps, may be observable.

The motion away from us causes a stretching of the waves, just as we discussed for the redshift expansion of the universe. And when the star comes towards us, the waves will be compressed into shorter wavelength, **bluer** light, and a blueshift will be detected in the compressed wavelength for the light reaching us from the star. The same precision-type of spectrometer equipment may be used as we saw in Chapter 6 to measure the motions of galaxies

The discovery of the **first planet** of another star was by the latter method, through spectrometry, and announced with great fanfare in 1993. By 1999, we knew of 30 systems to have a star with a companion, and 21 stars without. By 2003, 100 planetary systems had been discovered, 137 in April 2005, and the hunt continues with a variety of methods, recently also on spacecraft[1]. Multiple systems do appear in the observations, the ones that have more than one planet, but it is difficult to sort out the multiple signals.

Only a few of the systems discovered thus far seem similar to that of Sun and Earth[1,2]. A planet like Jupiter is much more massive than Earth so that one finds the effects of large planets primarily in these investigations. While most of the planets found thus far have masses like that of **Jupiter**, they do not occur at Jupiter's distance from the sun. They are much closer in with their orbits. Most searches have not had a long enough time-span yet, in order to cover periods of 12 years or longer for the planet at large distance from its star. There is also a suspicion that some of the objects found may actually be small stars, "brown dwarfs", large for a planet, small for a star. All of these are intriguing discoveries, so different from what was expected.

A next iteration of observing will be over a wide range of wavelengths on **spacecraft** in order to avoid interference from the terrestrial molecular signatures[1]. Looking for oxygen with spectroscopic equipment, for instance, seems logical since one believes that the oxygen in the terrestrial atmosphere came about through the interaction with life forms. Kepler and Darwin and the Terrestrial Planet Finder may do that from space, joining the Space

Interferometry Mission in looking for new planets, in a possible collaboration of the European Space Agency (ESA) and NASA. Our expectation for finding planets of other stars is coming true and that gives confidence in the similar prediction, that there is life out there, perhaps including intelligent life.

15.2 Are we Alone in the Multiverse?

One understands the evolution on our planet Earth so well, seeming so repeatable, that we expect to observe complex forms of evolution on any planet with suitable conditions. However, exactly what these conditions are and how far these can deviate from those on Earth, are topics of lively debate[2]. It is emerging as a new discipline of science, **astrobiology**. The detection of the first amino acid in interstellar space, glycine (Table 6.1), was a milestone in this new field (Table 6.1). The 12-meter telescope mentioned in that table, with which the discovery was made, is now mostly dedicated to astrobiology studies under the auspices of the newly instituted **Arizona Radio Observatory.**

The conclusion that there are deep terrestrial layers with basic biogenetic material (Sec. 11.1) is especially important for the predictions because the same mechanisms of accretion that occurred on Earth may have brought organic molecules from the interstellar medium also to the planets of other stars. This sentence brings us into the heart of the debate, because the biogenetic material emerges on the surface of the earth through underwater plumes caused by plate tectonics; water is essential, the new planet must have abundant water, and that may be a rare condition. If that condition were satisfied, it would be intriguing to realize that the composition of the biogenetic material **might be the same** on the new planet as on ours. There are long stages of mixing and perfecting for the molecules in the interstellar medium, and telescopic observations show exactly the same spectra for each molecule when observed in different interstellar clouds (Sec. 6.1).

To show how specific considerations enter in, a numerical prediction within our galaxy is exercised with **Drake's Equation**, which was devised as long ago as 1961 by astronomer Frank Drake[3]. As you read this, make your own guesses. This field has a contrast of optimists (life everywhere), and pessimists (none other than ours), and their estimates range widely. Let us hear from the optimists first. It is for **N**, the **number of civilizations in our galaxy**,

$$N = R_s \times f_p \times n_e \times f_l \times f_i \times f_c \times L .$$

R_s is the rate at which stars like our sun form in the galaxy, and it is derived from what is generally known about the topic. Drake and others have made the following estimates. R_s = 10 per year, and all stars are fairly well known to have planets so that fraction f_p = 1. The number, actually the fraction, among these of planets with an environment suitable for life is according to them n_e = 1, and so is the fraction of habitable planets on which life actually exists, f_l = 1. The fraction of those life forms that evolve into intelligent species, f_i = 0.01. The fraction of those species that develop adequate technology and then choose to send messages out into space, f_c = 0.1. So we see that scientists like Drake, who are in the business of searching for signals from life elsewhere, derive numbers like N = 0.01 L, where L is the lifetime of such a technology-advanced civilization.

Quite different values are derived for the above terms **by the pessimists**, because so much is still uncertain[2,3]. The values may have to come down appreciably when taking into account the unique importance of the moon stabilizing the earth's rotation axis in space through tidal effects. If those were missing, the movement of the earth's rotation axis might be erratic to such an extent that the conditions for life on Earth would be too drastically unstable over relatively short times. Combined with the small chance of a glancing collision on early Earth to produce the moon, it brings the number of potentially suitable planets down steeply. Next, is the question whether a moon of the earth is likely, or is it rare, near 10^{-4} perhaps? Additional requirements for planetary astrobiology are of plate tectonics and large Jupiter-type planets as having been important in the evolution of life. Would N = 10^{-8} L then be a better guess? That number differs so much from the previous 10^{-2} that the predictions seem quite uncertain.

This is all within a galaxy, but the question is for the multiverse in the title of this section. There are at least a billion galaxies, 10^9, in our universe, and perhaps an infinite number of universes in the multiverse, so we are **probably not alone in the multiverse**, even if one is a pessimist, also regarding L being low. However, I know of pessimists regarding the multiverse model at the end of Sec. 6.6 being real, so there is their chance. The L term is an unknown we still need to discuss.

15.3 How Long will an Intelligent Society Survive?

Now we must consider the lifetime of an advanced society, L. Beyond the span of natural life cycles, we need to take into account the chance of such a society destroying itself. We ourselves are a miserable example. In 1962 our world came close to a **nuclear** war that would have killed or mutated most of humanity[4]. The way our world is arming itself now, one could not possibly be optimistic and trust our species over the millennia, especially with our recent renewal in nuclear devices and arms race (Sec. 13.1). What is the value of our education, the worldview of this book, and its optimism that intelligent life might clean up its acts? On any planets of other stars there may have been setbacks like on ours, such as epidemics, asteroid impacts and other disasters, but **intelligent** life finds solutions.

As for the value of **L**, make your own choice. I choose it following the reasoning in this book. A species that is able and wants to communicate with others must be intelligent and confident enough to maintain itself. Such an advanced species **will not destroy itself** through overpopulation, nuclear or other warfare, let alone by its own pollution. If this reasoning were true, **L** would be large. The largest it can be is the lifetime of the sun. Even for its maximum $L = 10^9$ years, but using the pessimists' $N = 10^{-8} L$, **N** is only 10 for a galaxy (of which there are billions of suitable ones in the universe). Note that the optimists with their $N = 0.01 L$ would show $N = 10^7$, ten million societies in a galaxy, a crowded universe. This is not to say that all parts of our galaxy are suitable for life on planets, the galactic center for instance - I do not know how to estimate that, and the following paragraph avoids that problem anyway.

For visiting and communicating, contact seems realistic only over distances of not more than **50 lightyears**, 50 ly. Who would want to wait more than a lifetime for a reply to a signal sent? Let us reverse Drake's reasoning of the rate at which stars like our sun form in the galaxy. All we need to do is consider by what factor our little sphere, of 50 ly radius, is smaller than our galaxy, which has 4×10^{12} cubic lightyears (ly^3) because its radius is 50,000 ly and thickness about 500 ly; the answer is a factor of 10^7 times smaller.

For the optimistic Drake reasoning, without taking the moon and Jupiter into account, and the most optimistic $L = 10^9$ years, there would be one civilization in the volume with radius of 50 lightyears. And yes, **we are here**. But to be fair to the pessimists, we should conclude that the chance of finding advanced life elsewhere is a delightfully open question at this time; the optimists and pessimists will have to continue to be just that for a while longer. We move to the question if any spacecraft of the optimists' friends, or pessimists' aliens, have been seen yet, with or without living occupants?

15.4 Unidentified Flying Objects (UFOs)

"Are other societies studying us?" is the question that keeps the UFO fascination alive. Let us examine the observations and then estimate the likelihood for their visits. Reports of unidentified flying objects come from sightings and a few photographs of strange objects that seem to most viewers to be supernatural and of extra-terrestrial origin. The observations are generally of transient events, many of poor quality and poor viewing conditions explain most of them, or atmospheric phenomena, or college students' pranks, or by some outright **fraud** in this business. The latter bring fantastic tales of ancient societies that landed on Earth and left landing strips in Peru and pyramids in Egypt. Their books sell well, but the fantasies are those of the authors and not substantiated by proper scientists as far as I know. A few percent of the UFO sightings, however, seem unexplained.

I recommend to everyone to look out of windows of airplanes, having reserved a seat on the side where the sun is not. You might even see a rainbow where raindrops are present in the atmosphere below, and the rainbow will be seen upside-down from what we are used to. Much more often is seen exactly opposite to the sun, the **opposition effect**^{5}, a small faintly shining area, entertaining as it follows us everywhere and having different brightness and appearance depending on the nature of clouds or other reflecting objects. Wherever corner-cube paint is used, as for highway signs, the sun will be flashing back at you, and rear lights of motor vehicles will do the same.

As for UFOs, I saw one on a flight over Europe when there were many haze layers that can be amazingly thin. Some simple thought can help us conclude what it really was. Ahead of our plane, I could see the thin and irregular haze layers we would soon fly over, and then the shadow of our plane indeed spooked around us, sometimes approaching sometimes receding. However, the pilots of that 1989 flight might have reported it as a **UFO sighting** with conviction, because the haze layers were hardly discernable, and fuzzy enough to make the shape of the plane different and spookier. I had been looking and checking them intently for a long time, which is something pilots can rarely do because they are busy within their cockpit and in nearly continuous communication with ground controllers^6.

In order to estimate the **likelihood of a UFO visit**, we can use the reasoning in the previous section from which we concluded that within 50 lightyears distance from us, the chance of finding advanced societies is quite uncertain but small, lying between one and one-in-a-million, 10^{-6}. The society would have to live on a planet, of course, and no planet has been found nearby, assuming that all stars within 50 lightyears have been observed for the purpose, which is probably true; results of 'not finding a planet' are not always reported. We know how to accelerate spacecraft gravitationally with repeated flybys of planets to speeds of about 30 km/sec. With ion propulsion or nuclear techniques we might reach perhaps even 3,000 km/sec, one hundredth of the speed of light. We are limited by our own stage of development, of course, and another society may be much more advanced, its space travel faster. So, for argument's sake, let it be one tenth of the speed of light. To travel the distance of 50 lightyears, it would then still take 500 years of one-way travel, through darkness, and successive generations would have to be bred on board for no other purpose then making this flight. Who would volunteer for such a mission? Who would fund such a mission?

There may be another reason that not a single manned UFO has yet come as a probe from another society, and this is that there is no coherence in the UFO observations. After such a long and costly trip, the explorers would not make furtive visits that do not seem connected or even similar, and separated by years. Following such a journey, it seems likely that the travelers would land, explore, and colonize as Columbus did, fearlessly because it would be quickly established that they were the proven, powerful, and superior beings. **The chance of a UFO with naturally living intelligent life on board seems negligibly small.**

A **robotic probe** would be more likely. For comparison, one might ask what our space program would look like to a Martian if not furtive and separated by years, while it actually is consistent with thorough exploration through remote sensing. In fact, as a planetary scientist, I would like to propose such a mission to explore the interplanetary and interstellar media both, through the Oort Cloud and then on towards a planet of another stars. In summary of this section, we should keep **an open mind** on the UFO issue where robotic probes are concerned. With that as encouragement, let us inspect the status of sending signals to and receiving signals from our friends.

15.5 Searches for Extraterrestrial Intelligence

An interstellar **message at the speed of light** is much faster and easier to execute than an interstellar voyage. It is, however, not obvious to everyone that humanity should search for intelligence elsewhere and thereby attracting attention to our presence here. The U. S. Congress discontinued all funding for making radio searches of extraterrestrial intelligence in the early 1990s - private donors have taken the SETI support over.

A considerable amount of searching has already been done, at a variety of radio telescopes. One reason that no signal from another society is received may be that we do not know when and where, and at what wavelengths their signals are transmitted. They may well be sending intermittently, over a narrow range of wavelengths, or by using some technique that we do not understand. We must make our searches soon, since in the future they may be impossible because of **interference** from our multitudes of TV, radio, and communication channels; we have a similar problem for radio astronomy in keeping suitable channels free from human interference; these problems are discussed at planning meetings for international agreements to reserve certain wavelengths for science. There are yet other restrictions. There is natural radiation from various sources of galactic emissions, mostly at wavelengths longer than 50 cm, and on the other, shorter side, atmospheric absorption occurs and blocks transmission at wavelengths below about 3 cm. The listening window is then rather restricted between about 3 and 50 cm, but that is also the popular region for communication channels on Earth.

The largest proposed project was CYCLOPS[3]. It was to be an **ensemble** of many large radio telescopes all pointed at the same object for either receiving or transmitting signals. CYCLOPS would have had an effective aperture of 3,160 m, compared to the presently largest optical telescope in the 10 - 12 meter class, and it might have received signals from distances as great as a thousand lightyears, based on estimates for the power at which signals might be sent from out yonder. However, the cost in 1971 was to be about 600 million dollars per year, and no such funding is available. Private and corporate donors have funded not only SETI searches with multi-wavelength receivers at existing telescopes, but now also a CYCLOPS-type of project that will eventually have as many as 350 inexpensive antennas of 6-meter aperture[7].

The **evolution of engineering** makes the situation more complicated. CYCLOPS was to look for signals from other planets including their "side lobes" of their transmissions, as we understood communications in the 1960s. For example, the traffic on Earth would be detectable, albeit weakly, out to a distance of about 50 lightyears by the noise from our own radio transmissions. But since that time our own side lobes have been minimized for greater efficiency, and an advanced society would have done that too, so that one now looks for signals beamed straight at us. Are they doing that, sending signals to us? We are not sending signals to them, other than a few exceptions, notably some messages by Frank Drake in 1974, and from a Ukrainian radio observatory since that time.

If SETI finds no signal, what does that mean? Would failure to locate a signal mean that there is no other civilized society nearby, or could there be a **technical reason**? Here on Earth there are already modernized communication techniques, such as with laser beams, optical light in narrow beams. Another advanced society would also be using their most powerful and advanced technique to beam at us, and we might have no inkling of what that would be. If the formation of their star and its planets had begun appreciably earlier, the leading species will be way ahead of us, they will have advanced rapidly beyond us, and we would not understand them. However, there would be little doubt about the language they would use or we would use.

15.6 How would we Communicate?

This seems for us at this time an ultimate question, whereby we prepare for learning from other societies about them and their evolution, and it is also ultimate in that we need all we have learned in this book to answer the question. The most direct communication with other civilizations would of course be to **visit** each other, as we discussed before, and there is no result with that yet. Next best is to **send a written message**, and we have done that on four spacecraft, aboard two Pioneers and two Voyagers. However, the chance of getting close enough to another civilization, for them to collect our present by fetching the spacecraft, seems negligibly small because there is just too much space between the stars. We considered aiming at any star near the directions Pioneer 10 and 11 were moving after the end of their scientific missions in the solar system, but it was not feasible with the small amount of hydrazine control fuel left.

Even if the chance of delivery is remote, it is an intriguing exercise for us to think through how to communicate with others. The "language" will have to be based on science and science they have, or they could not have caught our spacecraft and received our announcements of its arrival in signals from Earth. It is at least a challenge for **our** imagination to design messages to remote and totally unknown civilizations.

The first exercise, by Carl Sagan, was a plaque with an inscribed message to the leading species that will have followed us, evolved on Earth eight million years from now. The occasion for this exercise was the LAser GEOdynamics Satellite, **LAGEOS**, a passive spacecraft (no maneuvering) to which laser pulses are sent from Earth. From the received reflection of the pulses one can measure the round trip time and derive the precise distance (the radar principle is described in Sec. 3.4). The goal is to precisely measure changes of the Earth's surface and its rotation axis. The first launch was by the U.S. in 1976, and LAGEOS II was launched by Italy in 1992. In Earth orbit, the lifetimes of the high spacecraft are about 8 million years, after which the friction with and slow drag even by the very upper layers of the atmosphere will cause them to fall destructively back to Earth.

In preparation for our exercise in deciphering a space image, below, we should be familiar with the **binary code** in electronic circuitry, for the other society must know that too (8 million years from now on Earth, or anytime somewhere on a planet of another star). It is the simplest of communication systems, requiring only two digits: 1 or 0, or as it is in electronics, "on" or "off." Just as each place in our decimal numbering system represents a power of ten, each place in the binary system is a power of two. That is usually written from right to left, and we shall see why. The first place is 1 (considered as 2^0, which is defined as 1), the second 2 (2^1), the third 4 (2^2), the fourth 8 (2^3), etc. For example, the binary notation 1010 read from right to left means that there are a 2 and an 8 which add up to 10, so the resulting number is 10 since the first and the third position are left open and do not contribute.

Now we are ready to look at the message for the two **Pioneer** spacecraft launched in the 1970s to fly by Jupiter and Saturn; they have traveled far and are leaving the solar system. Frank Drake and Carl Sagan designed the plaque of Fig. 15.1. Pretend to be a member of the other civilization and try to decipher the figure. If you need help, here it is. It seems logical to communicate by using mathematical and physical principles, for advanced societies know them even if they do not know each other's parochial languages. The **hydrogen molecule**, so basic in the universe, will be quite familiar to our extraterrestrial colleagues; it shows with two atoms in the upper left of the figure. Its electron waves have a property in quantum mechanics that we call **spin**. It is not exactly a rotating spin but it reminds us of a spinning top and so it is probably to them - don't you think their children play with tops? The electron of the hydrogen flips its spin property, like a top flipping upside down, and back again (which a top cannot do, but the electron performs something like that). So, the flip shows with two opposite directions. Look carefully for the small end blobs of the vertical lines: for the atom on the right, the electron spins in the same sense as the nucleus, while on the left its electron has the opposite state. If our colleagues are advanced enough to have been able to catch and collect this spacecraft in recognizable configuration, they will readily recognize the transition from one spin to the other, and know that this produces radiation of a definite wavelength, which is 21.11 in our centimeters. That unit of

21.11 cm occurs in the message as a unit, a standard for the measurement of length. If they are really advanced, they may have that as their standard, instead of our having the rather arbitrary standard of a meter-long rod at a certain temperature that is kept somewhere in Paris.

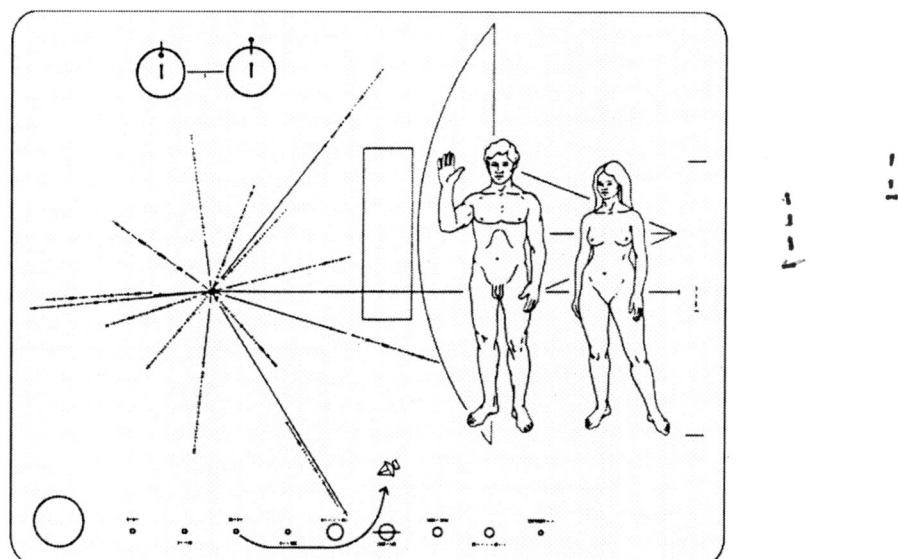

Fig. 15.1. The plaque on Pioneers 10 and 11.

A binary number gives the woman's height and now it is read from top to bottom. There are no contributions in the first, second and third places, but there is a mark in the fourth place, so the number is 8. They have the spacecraft shown in the background, they will understand that the height of the woman is 8 times the above wavelength. We would say that her height is 8 x 21.11 centimeters, but note that the French call and pronounce it already different, 21.11 "centimètres", even though it is the same unit. They may have the spacecraft in more or less undamaged condition (Fig. 15.2), they will measure its dimensions, connect that with the indicated human height of 8 times **whatever-they-call-it** for that hydrogen flip, and they will surely put all these clues together.

Our intelligent colleagues will have **a wonderful time** studying the human figures. They will understand the idea of male and female if I am right in my conclusion that their own life forms developed the same as ours, from the same molecules and the same early origins and universal evolution. They may not have ten fingers, or a decimal system. However, they will tolerantly conclude that our human beings use a numeric system, in addition to the binary, based on five or ten, because of those five fingers on each hand.

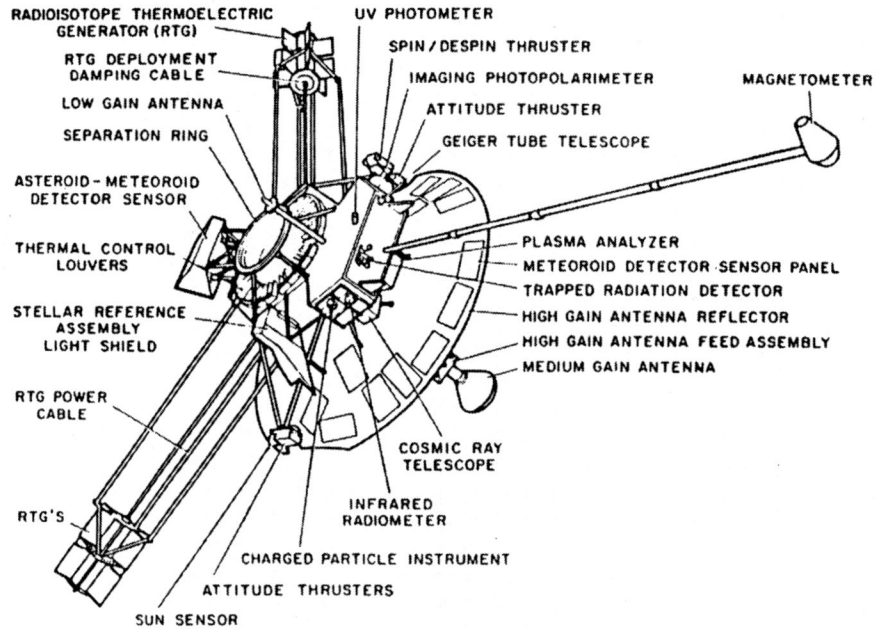

Fig. 15.2. The Pioneer 10 and 11 spacecraft.

The spacecraft sketch is again in the bottom of Fig. 15.1, and they will conclude that those lower sketches are of our **solar system**, the large circle is our "sun", with nine "planets". The spacecraft had departed from the third one out and passed by the fifth. The distances of each planet from the sun appear in a binary code.

The star-like figure on the left indicates the directions in which we observe pulsars. The line to the right, extending to just beyond the woman, indicates our distance of 26,000 lightyears to the center of our galaxy, which is their galaxy too. Using that distance as a unit, because our colleagues know **approximately** where we are to begin with, at about the same distance from the galactic center as they are, the lengths of the other lines show our distances to these pulsars. Beings advanced enough to have been able to intercept the Pioneer spacecraft, they must be familiar with pulsars, and they will determine from the intercept of the pulsar positions more precisely **where our sun is** in the galaxy.

Finally, there is the determination of our approximate **year of launch**. The binary code along the lines of the pulsars indicates the periods of pulsation for their identification, and they again will know those periods. What is fascinating here is that the pulsar periods slow down with time. The extra-terrestrials know the pulsar periods at the time they find the spacecraft and the rate at which the periods slow down, so they can figure out the indicated periods near the time of our launch.

On the two Voyager spacecraft, a much more detailed data set went into space, designed by a team working with Carl Sagan. Each set is on a phonograph record, with instructions for playing it, plus a cartridge and a needle. The historians of the other civilization will be interested in our obsolete techniques (we ourselves have already progressed to CD-ROM for such recording, to DVD, and to direct digital recording in Mp3 format). There are sketches on the gold-plated recording similar to those on the previous Pioneer spacecraft, but there also are 115 coded images of **scenes** on Earth. The recording has greetings spoken in some 55 languages, and a fascinating collection of **sounds** of the earth, ranging from a thunderstorm to opera music to a kiss. No one may ever listen to it, but it was a jolly exercise to conceive it, and hard work too - Carl duly reported that they had to practice the kiss many times.

16. WHAT IS NEXT FOR PLANET EARTH'S SOCIETY?

16.1 Coming Together

This book overviewed our contemporary world with the latest observations regarding universes and humanity, their origins and future. It quoted the greatest teachers to learn the **ethereal** aspects of universe including humanity, which seemed at first so unbelievable but we learned where these aspects came from. We studied space, waves, particles, and gravitational space-curvature, how they interact and thereby connect everything. An appreciation of evolution and quantum physics was developed to understand a little better the wonders and mysteries of nature, and of life and death.

For this chapter, there are a few special topics to fill in between what was discussed before, to round it off and make it all come together. For example, 26 delegates from 5 continents assembled an **Earth Charter** as a type of Constitution of the World, with 72 Principles for a Sustainable Future. Their main areas are 1) Respect and Care for the Community of Life, 2) Ecological Integrity, 3) Social and Economic Justice, and 4) Nonviolence and Peace. A brochure with careful text is available[1], but what seems missing in the Charter is a consultation with the Muslim world, and there is no common guideline for execution of the principles, for finding solutions to modern problems as we tried to do.

Planet Earth's society is at a crucial point in its history. The most urgent global crises seem to be **weapons sales, nuclear warfare, global warming, poverty, and asteroid impact.** The latter is fairly well under control, even though the danger cannot be eliminated completely because of continuous supply asteroids from their main belt and comets from the Oort cloud. The surveying must continue, and engineering for mitigation must be prepared. The other crises are predicted to occur on a timescale of a decade or two.

The most powerful country to do anything about the first four of these global crises, the U.S., has declined in international reputation, it is in fact driving the first two, and has largely ignored the next two. However, it has led the world against the asteroid danger, and we may hold that up as an example. The U.S. is a democracy, it is the American people who decide and can make a change. Are the problems clear? Let us look at them in this chapter again, in sharpened views. Here are two vignettes of the U.S., for where it came from and for where it may be headed.

16.2 The Rise and Decline of Societies

We begin with an evaluation of the current position of the United States, by a Chairman for the Union of Concerned Scientists[2]:

> The United States is in a position that no state has held since the Roman Empire...It alone is able to project military Power globally...It has the largest economy....It is home to the majority of the most productive research universities and most successful high-technology firms, and thus attracts a disproportionate fraction of the world's scientific talent ...

Being an immigrant in the U.S., I am intrigued how the above situation came about and what immigration might have had to do with it. The mixing of peoples may be one reason, among others, why the U.S. is so successful. America's immigrants have come and are still coming from all directions and, by inter-marriage, they produce the greatest **variety of chromosomes** that has ever been mixed in time and place on Earth. This brings a wealth of choice of capabilities and characteristics for the conceived generations and thereby for the power and enterprise of the Nation. The opposite effect is seen where extreme isolation causes inbreeding, which may bring deleterious defects due to the restriction of choice among the chromosomes of the related parents.

Mixing of a large variety of people's strains seems to bring about the strength of the U.S. This may be a repeat of what happened in Western Europe. We can then understand better how European traders and military in small sailing ships were able to spread all over the world. We can also understand the roots of globalization. We should understand next how societies come and go.

Much has been written about the logical patterns in the **rise and fall** of empires and societies; they may be summarized as follows. First, original settlers built their home and society in the new land, and additional migrants came from leaving or fleeing their land of birth for various reasons, but all of them had to be tough and frugal people occupied with safeguarding their families and their survival.

The settlers were usually tolerant of newcomers because they knew of the **hard times** they, their parents, or grandparents experienced in building their new society, and they were open to the addition of new labor and ideas. Patterns in farming and enterprise were developing.

The immigrants practiced their own forms of cultural and religious behavior, but **not fanatically**, as they were often too busy to focus on their differences and they valued each other's support. The new society would engage in development of new resources, natural and manmade, and a pursuit and defense of freedom. When the economy began to prosper, the arts and sciences appeared as well as could be afforded. These success stories would bring pride and respect, and an influx of additional people so that the mixing of chromosomes and accomplishments came into full swing.

To everything comes an end. Prosperity's personal wealth makes the society members gradually **more possessive** of their luxuries. They become more protective of their own interests, making friendships that sustain these interests, and as a result they become more reluctant to accept newcomers and foreigners. Outsourcing of their own inventions results in cheaper products, and the globalizing of their industry and warfare protects these bigger profits. There is less interest in international agreements that might be costly, nor in helping the poor, especially of faraway places. Gradually a decline shows itself as a playing up to the wealthy and in a fundamentalism geared towards keeping the status quo. There may be local and temporary exceptions and reversals giving the appearance of no decline, and prosperous economies may bring another leap into the future.

For the United States, we reported a decline of international respect and its pattern of refusal to ratify or maintain existing international agreements of the World Health Assembly, the seabed and mine-ban treaties, and attempts at tackling global warming and chemical and biological warfare. The Nation may be doing well internally and not notice that its currency is devaluating and that it is deeply and increasingly **in debt**, while depending on lenders to set the rate of interest, lenders to whom American values are foreign and possibly not sympathetic[3].

Tumultuous times and turning points in history have occurred before, and people's resilience has saved the day before. A classical example of such a day was 10 May 1940 for the United Kingdom, Western Europe and ultimately the whole world, when Hitler was about to win World War II, but **Winston Churchill** declared that England would stand the onslaught, in blood, sweat and tears.

Does our ethic of global responsibility have anything realistic to offer in the present world situation and possible turning point? We are living close to each other in a strongly interacting and weaponized world. The greatest danger is that the present situation leads to **nuclear warfare**, which would bring radioactive fall out and irrevocable irradiation into the vilest mutations. The generation of Sakharov, Eisenhower, and Khrushchev was keenly aware of that impossible risk, but a new generation may have forgotten. Chemical and biological weapons are also available. Usage of such weapons, for whatever reason, would cause a **futile ending** of humanity, which could have tackled with its brain rather than brawn its poverty, excess of weapons, and unbalance of power at much lesser expense.

Eisenhower warned us already in 1961 that we must learn to get along. He was speaking to Americans for they have the dual power to destroy or save the world. They have the awesome responsibility to keep themselves informed about their international problems and realities. They must educate themselves to be able to use their electoral system for selecting leaders who understand the global situation. **The current situation is as urgent as it was for Churchill.** Let us look at this crisis from another angle.

16.3 Survival of the Thoughtful

Many Americans believe that it is natural for the influence of the U. S. in the world to be increased, that it should stem from more power and if necessary greater application of force[4]. This misunderstanding sometimes emerges in my classes too, that evolution means the survival of the overpowering forceful. That is indeed the case for animals with the simplest brains, but evolution brought increasingly more complex and capable brains such that problems arose but also that survival moved away from the strongest, towards the usage of more intricate means of defense and offense in human relations, business, and politics. We see **the survival of the thoughtful,** of the species that rely on their brain to survive. It is now up to our society, to survive, or not. Please, do not think this is rhetoric, because those of us who lived through the Cold War – and its frightening Cuba Crisis date of October 28, 1962 – know how close we came to extinction (Sec. 15.3). Let us recall what happened.

Physicist Andrei Sakharov designed the hydrogen bomb for the Soviet Union, but then became a dissident to the regime because **he had learned** that nuclear weapons violate the possibility of life's survival[5]. They have no rightful place on Earth. There is no valid excuse for developing them, because of their permanently destructive effect on life. However, only thoughtful, properly educated people can understand this. This realization has slipped by the awareness of populations such that there are now perhaps a dozen nations that have or soon will have nuclear weapons. It will take the greatest demands on thoughtfulness to solve this problem before it is too late. I am not referring to U.N. resolutions for sanctions on nations developing such new weaponry. We learned on October 28, 1962 that the largest Nations are the most dangerous in this respect, and Sec. 13.1 shows why the U.S.A. is now the most dangerous of all. We must disarm, the U.S.A. included, or we shall perish. Time is running out, as it is running out on global warming, both with time scales of only a decade or two, and both can be resolved by thought, rather than by continued greed, power and negligence.

We discussed with Fig. 11.7 how the brain evolved in capability. Even the rabbit though, with its small mammalian brain, propagates its species by warning its community of impeding danger by slapping of a hind paw on the ground, in spite of its own individual risk as the slapping betrays its position to the predator. This is **altruism,** unselfish devotion to others (Fr. autrui, other people). The latest addition to the evolution of the brains was the **cerebral cortex** of humans, which brought the capability of sophisticated planning for the cleverest schemes and courses ahead; it brought the tool for advanced education. We all know examples of clever successes in science, business, engineering, and human relations. Evolution developed the future by random choices of what was to come next, until it made tools so complex that they began to mate and thus bring about the variations that are evident within the species. The human cerebral cortex is now in control of the future. Let us find a guideline for it to solve global problems.

16.4 The Evolution Guideline for Action

Continuing on the theme of evolution and education, it brings its guideline for taking actions by **continuing the course** of successful evolution. Modern business and military commanders know that any time the competition or enemy is on the run, it is not the time to stop but to proceed. Another way to explain and remember the guideline is as if we are in a boat on a fast-running stream, not to row against, but to row with the current. Dangerous rapids have "wedges", where the experienced boatmen recognize that the rocks are the least destructive, so that they emerge intact and ready for new initiatives and ventures. It is common sense.

The guideline works over a wide range of **examples** and is therefore applicable to any problem or legislation. We begin with a simple example from our daily observation of traffic rules: drive recklessly and end up in a hospital, or worse. The next one goes a little deeper. Since about 1995, there has been a trend towards much larger automobiles than necessary, serving only personal pride and pleasure at the expense of forcing society to live in a more crowded and polluted world.

The third example is broader yet in that we have noticed that the U.S. refrains from ratifying international treaties and protocols. In daily life, it is like a family wanting to visit and deal with the neighbors but not wanting to share the burdens of the neighborhood. The guideline points towards providing a proper environment for growth and development to the neighborhood, for acting as good neighbors, rather than selfishly destructive loners. But now the "neighbors" are nations, and the needs are greater.

16.5 Integrity among Nations

The evolution of life brought an increase of interaction in ever widening circles. Humanity evolved together with its environment from nomadic tribal units to villages and cities and to states and nations. The **federation** that is the United States is an example of States effectively working together while keeping local matters in their own hands, while their foreign interactions are performed by a united representation. A modern case is the growing combination into a united Europe, in spite of having been at each other's throats in recent centuries. Could and would Europe[6] take over leadership from the U.S., and if so, to what extent? The Euro is already becoming the currency to invest in, as it used to be in the Dollar[6]. China, India, and Russia are also up and coming.

The time for closer international collaboration appears upon us because the world is rapidly becoming more interdependent, with globalization booming and greater hazards looming. The opposite view comes from **patriotic feelings** but also from political and commercial pressure on using military force. It looks unlikely that a candidate for the

Presidency of the United States could campaign successfully on letting authority of the country and its citizens pass into foreign hands.

The survival of the present world order as well as of the U.S. may be at stake, such that respectful international relations are a natural and essential precaution. For example, letting terrorist hunting be the detective policing of **Interpol**[7], a confidential and highly effective organization since 1923 (we shall discuss Interpol in other contexts below). A professional review of the goals of the U.S. within its global community seems overdue. Such a scholarly examination could use the **World Scientists' Warning to Humanity** issued in 1992 by 1,700 scholars known for their accomplishment in a variety of expertise[8]. The warning was updated after September 11, 2001, by 105 Nobel Laureates[8]:

> It is time to turn our backs on the unilateral search for security, in which we seek to shelter behind walls. Instead, we must persist in the quest for united action to counter both global warming and a weaponized world...
>
> Some of the needed legal instruments are already at hand, such as the Anti-Ballistic Missile Treaty, the Convention on Climate Change, the Strategic Arms Reduction Treaties and the Comprehensive Test Ban Treaty. As concerned citizens, we urge all governments to commit to these goals that constitute steps on the way to replacement of war by law.

The opposite view prevailed, with the U.S. withdrawing from these three organizations. Encouragement of international collaboration was already in President Eisenhower's speech of 1961 (Sec. 13.3); the heading of this section comes from it:

> Throughout America's adventure in free government, our basic purposes have been to keep the peace; to foster progress in human achievement, and to enhance liberty, dignity and integrity among people and among nations...
>
> Any failure traceable to arrogance, or our lack of comprehension or readiness to sacrifice would inflict upon us grievous hurt both at home and abroad. America seems to have the awesome responsibility to lead the world, and might be allowed by the world to do so thoughtfully rather than violently, to raise the world and its poor economically upwards along with its wealthy.

Such a comprehension or tolerance by the U.S. will be rewarded by **opportunities** in new industry, enterprise, exchange of goods, to all, including Americans. In consideration of the above parable of being neighbors, the leadership has to come from within the neighborhood, constructed upon respect among its members, no matter how different and basically hostile they may be. It will take the wisdom of the electorate to select leaders who will do that.

The opposite view, held by those who will not vote for or remain bound by international Treaties, is that the **status quo** seems just fine, America's internal economy is doing well while abroad it can afford to act unilaterally, and that this is the opinion of the majority of the constituents they represent. Let us examine how useful the education from opposite views can be.

16.6. Learning from Opponents

This section reports some useful experience with terrorism during and immediately following WWII when I had an opportunity to observe people judged by some as terrorists. **Terrorism** as described in dictionaries is the systematic use of violence, intimidation, and overwhelming fear to achieve an end. Often it is the poor man's warfare, a variation on being partisans or guerillas when their enemy is locally more powerful than they are[9]. Regular warfare too brings such fear, has notorious cases of criminality, and kills large numbers of civilians, especially in WWII; a result is usually that volunteers emerge, with various backgrounds and causes for which they would stake their lives[9]. Winston Churchill had initiated an organization to use such people for stimulating uprisings, "to set Europe ablaze," by the Special Operations Executive (SOE)[10]. A typical example of recruits was Sep Postma, a young Dutchman and sensitive idealist who could see the danger of Hitler, who therefore escaped to England, parachuted back into occupied Holland in 1944 after half a year of training in England as an agent[11]. He was caught and killed by Dutch traitor troops. His body and those of about nine members of the Resistance, were put on display in the streets of Apeldoorn with signs: **"TERRORIST"**. But there was no shortage of volunteers for SOE, I was among them in the Resistance, escaping from Holland and being sent back like Postma, with radio operator Wouter Pleysier, and a second time with Ferdinand Stuvel.

But then, soon after the end of the fighting in Europe, I found myself on the other side in the Dutch East Indies, believing at first that we were hunting terrorists but quickly discovering that we were mostly fighting

freedom fighters who had no shortage of volunteers either and won their cause too. The opposite view was that of the Dutch government and of most of my colleagues in Indonesia, that we were primarily fighting terrorism. The situation was as confusing as Vietnam was for the U.S. in the 1970s.

No one disputes the rights of the Germans to pursue Churchill's SOE, or of the Americans to pursue the attackers of September 11. There are, however, two examples on the pursuit of partisans and agents who operate in civilian clothes. First, **Marshal Tito**'s partisans were effective in Yugoslavia even after the Germans sent in a million troops who operated in Army and Air Force formations, allowing the partisans to slip away from the turmoil in civilian clothes, while the military warfare killed many civilians, which brought even more volunteers with lifetime hatreds. On the other hand, a few experts **stealthily hunted** agents, as police detectives do that with city criminals and Interpol does internationally. The Germans succeeded in The Netherlands with only half a dozen of German experts to track and eliminate as many as 60 SOE agents and a large number of their Resistance supporters[10].

We learned at SOE to **study the opponents**, their causes and methods of fighting. Robert McNamara speaks of that as a lesson he had to learn as U.S. Secretary of Defense during the Cold War[12]. Such wisdom seems to be missing on the American side since the attack on the World Trade Center in New York in 1993, or one might have learned enough to prevent the September-11 attack. Removing the American garrison from the Holy City of Medina might have done it. There is this strange feeling in the U.S. that it is improper or unpatriotic to consider what makes terrorists do their deeds. Why do so many protesters[13] converge on the meetings of the World Trade Organization? The prestigious Journal, *Aviation Week* came out with a rare exception immediately following September 11, 2001, that the "... intensification of the terrorist threat is reckoned to be an integral part of the globalization of the world economy and international finance...[14]."

This book is for readers of all creeds and nations to consider all sides, to help find ways to live together on this intricately beautiful globe. We ought to see our volatile predicament in the larger context of how to survive in **an interdependent world**. One side's terrorist is usually the other side's freedom fighter. From my experience on both sides, I pose that we the people ought to make a sustained effort to diminish contempt, through education and diplomacy.

Let us begin to study the problem of poverty here in the context of terrorism so we can concentrate on poverty as a global problem in the next section. We may preach peace and ethics, but these teachings will be of interest to the poor only after the rich give them a chance to lift themselves out of their misery. A colleague in Ahmedabad connected terrorism and the chasm of rich and poor by posing a question to me soon after September 11, 2001, "How would you see Osama bin Laden if you were one of the many people in our streets **who own nothing**, nothing at all?" The answer to his question is clear when we see bin Laden's photo in the hovels of the poor and in demonstrations on the streets of Third World countries.

We must all consider my colleague's question, because large numbers of the world's people go hungry and seldom have clean water to drink. Do you realize what that often-heard observation means? They obtain their **drinking water** mostly from skimming sewage. When our daughter was little, I took her into the slum next to the Mumbai Airport. There she met two girls her age, well fed and dressed, but of low caste, not allowed at the local faucet even though such discrimination is now illegal. They collected the family's drinking water by lifting a cover stone and skimming the water from the sewer culvert of the majestic Centaur Hotel.

The 104 Nobel Laureates whom we quoted above made a warning regarding a new type of terrorism[8]:

> The most profound danger to world peace in the coming years will stem not from the irrational acts of states or individuals but from the legitimate demands of the world's dispossessed. .... If we then permit the devastating power of modern weaponry to spread through this combustible human landscape, we invite a conflagration that can engulf both rich and poor. The only hope for the future lies in co-operative international action, legitimized by democracy.

Their warning is for **street terrorism,** and it has a well-known source of stimulation. Vikram Sarabhai's television, which we have credited for having initiated birth control into the villages of India, has brought other insights as well, such as plans for leaving the village. Television is now available in all hamlets and slums - try then to imagine yourself being there: Would you not grow frustrated if you were hungry and despised and then saw daily showings on television of how splendidly the wealthy people live? If you were young but saw no hope for the future, would you not eventually rise to protest? As parents, would you not support their rebellion?

It thereby seems possible that the chasm of rich and poor in the world may bring terrorism on unimaginable scales. I am referring here to simple street terrorism with knives, stones and fire as weapons, but effective in stopping movement of the wealthy. I wonder about that when I ride in a rickshaw through the cities of India, how easy it would be to stop travel and commerce by westerners and their associates in street traffic. A summary expression in the intelligence communities for this eventuality is that **the streets of the East are hot**.

A locally informed background for this assessment is provided by David van Noortwijk, an experienced reporter in Jakarta, who wrote in October 2001 that understanding September 11 "... could perhaps help us to identify the real problems at hand, and better see the link between the hard line extremists and the average Muslim on the street anywhere in the Islamic world. Indonesia could be a useful barometer for the true state of affairs when America negotiates the dangerous waters ahead. There is in Indonesia a decidedly fervent and widespread wish for a more pronounced and conformist Islamic experience. Considering how moderate Indonesian Muslims used to be, why did Islam grow into the political force it is today? Time may show that we, the western strategic and capitalist alliance, are currently being presented with the bill for our past ... The fact that dangerous people like bin Laden and Saddam Hussein were previously supported by the U.S. is more than ironic; it's symptomatic. Some of these cases go back almost a century and they hardly seem relevant to us today, but we cannot ignore responsibility for the sins of our forefathers... No matter how hard it may seem, America and Europe have the responsibility to find a way for undoing past mistakes, and we must do so in modesty. For the enemy we're facing is among us alive in the world today. Deep down at its very core, I believe this is just another struggle between those who have and those who have not."

In summary of this section, my colleague in Ahmedabad and the reporter in Jakarta independently seem to show the change of bin Laden from opportunistic terrorist on September 11, to now being seen by many people as a heroic leader. The U.N. and nations individually offered advice and help after September 11 because of its terrorist nature and its human tragedy in New York, seen on television worldwide; some nations where many Muslims live were included at that time. **Interpol** could have quietly delivered the clique in Hamburg and others who had perpetrated the attack, and possibly bin Laden as well because he may have instigated it, and in any case he had approved it publicly. Instead of accepting that option, the U.S. made the most expensive, in lives and investments, global counterattacks, also seen on worldwide television, such that nearly a quarter of the world (Sec. 1.1) became its fanatic enemy and Osama bin Laden seen as a hero and leader, and, worst of all, it became felt by many as a religious conflict.

16.7 New Solutions for Poverty

The above indications are that a likely source of global conflict is the expanding chasm between rich and poor, the powerful and powerless. This was clear to see already in the 1990s when my Class Notes referred to that **chasm as a problem of the highest priority**, predicting that terrorists would attack the U.S. soon. Some people seem to believe that there always will be poverty, that it is incurable and unavoidable, that there are not resources enough to support everyone at proper levels, but Anne and Paul Ehrlich, who are prominent scholars of the future for the earth and its civilizations, are among the experts who disagree[15],

> It is entirely within the power of humanity to close the gap between rich and poor and to reduce the human population size to a level at which all people could lead a decent life without degrading the ecosphere. ... The great hope for civilization lies in the fact that people can recognize how the human predicament evolved and what changes need to be made to resolve it. No miracles, no outside intervention, and no new inventions are required. Human beings already have the power to preserve the Earth that everyone wants - they simply have to be willing to exercise it.

That was written in 1987, but little of the above has been attempted, not even during the affluent 1990s. Profit motives tend to preserve the status quo, even when new profits could be made in expansion to environmentally sensitive industries and the U.S. would be good at such invention and exploration[15]. Perhaps our global predicament is too foreign, too theoretical, as was the case for the asteroid hazard. But then the problem is not hopeless - there is a **lack of education**.

I introduced that lack and what to do about it during a memorable class discussion. There is a myriad of problems and aspects – where to begin? NASA stresses in proposals for a space mission, which usually address a myriad of problems and aspects, to sort out the **single-most important problem,** and then the others are usually solved together with the big one. "Well then," I asked the students, "in our complex global predicament, which is our single-most important problem?" The perceptive answer came back, that it is a **lack of respect** for fellow man and woman. For actions aimed at improving the future, this amounts to a **lack of love for the children.** Emotionally it is appealing to work for and raise funds for helping the children because they have the future, they are innocent and defenseless, and there are many parents in the world. Parents always say that they love and care for their children, but they must now do that in the global family; our love and care must reach the needy young of the world because they may become the ones to make or break our future.

The right for everyone to share in the environment and its resources emerges as a pillar for our ethic. No longer may anyone claim that the poor are poor because they are not smart or hardworking, nor may we keep them poor to slave for us, using child labor in sweatshops and cruelly violating their dignity in forced prostitution. Most of us who read this are fortunate to be far removed from these tortures, so it is easy to have a **lack of concern**. There is, however, one more reason why the chasm between rich and poor holds the single-most important cause of our many problems, namely that it will get rapidly more serious because of the television education described above. The whole world is on display to everyone, the terrorism attacks in the Middle East are in plain view as were the collapse of the Twin Towers in New York - the poor are no longer going to take their plight for granted and they do not need to do so anymore in this increasingly weaponized world. This world is in **a time of revolution** comparable to that of the invention of gunpowder, which brought feudalism down for good.

The opposing view to all this sees that child labor is efficient for supplying the rest of the world with **cheap products**; this feeling is rarely admitted openly, while in practice it is adhered to by a large fraction of the world's population. Are there enough vacancies, when so many people would be added to the job market? The answer is that they bring their own demand for new openings; the general level of economies would rise from the bottom up.

In spite of President Eisenhower's warning, **the industrial complex is here to stay**, with conglomerates doing business increasingly and in nearly every country, their foreign contacts and contracts expanding, and local employees fulfill supply and demand through foreign outsourcing. This has come forth from a natural expansion for trade and business into opportunities that had opened even before 1942. However, the environment has changed to one that the Nobel-Prize winners see as leading to a conflagration. To safeguard against that possibility as well as to be fair, the call is for **diversification,** which is doing business more the local way and to find ways and means to train the trainable poor, helping them to become a part of respectable fabrication and commerce. As they in turn become able to buy goods, the **economy is improved from the bottom up**.

There are **worldwide relief** efforts already on the way; I have even heard predictions that worldwide poverty will be a plague of the past merely some 50 years from now. Advanced nations have experience with debt relief, which the recipient countries put into education, health care, and development with their own contribution, provided there are no unacceptable strings and local interference attached. The United Nations has relief programs[16] as does the Carter Center[17], and The International Save the Children Alliance[18] works with adolescents, stipulating that they do not participate in armed hostilities. These are merely a few examples, while other organizations are easily surfed on the Web.

Collective action by all of us, supporting existing organizations, helps us assemble our individual convictions into combined action in national and international campaigns. An opposite circumstance is in duplication of organizations and fraudulent fundraising among donors who do not suspect that[19]. However, especially for the various needs of the children, a diversity of resources and supporters is in place and operation. Local funds are usually available to build or enlarge and operate the schools. There are contributions and participation by agencies and governments to cure the worst aspects of city slums, and there are slum projects with the involvement of private and corporate supporters[20].

We have reported the horrors of Bangkok where girls and boys are hauled by the truckload into cruel prostitution until they are too diseased to perform. This is an intolerable situation, but it is merely an example; Sec. 13.6 gives references to the astonishing extent of sex slavery today. A new philanthropic organization grew out of the passion of a college student to educate and provide better opportunities in life for the children of sex workers; it is called Achieving Sustainable Social Equality through Technology (**ASSET**)[21]. Its goal is to coach the children at a level of education and familiarity with technology, which will enable them to free themselves from being chained to that same profession.

A similar organization that coaches underprivileged children is **Akanksha** (translated from Sanskrit as hope and aspirations) whose reach in 2007 was for about 3,000 children in Mumbai and Pune, India[21]. A teacher fetches her or his children from their slum dwellings after school, and takes them for a few hours to a "Center" for 60 children each. Selected are the ones who are doing well in school, and have supporting parents. The training is for a 10-year curriculum to teach math and civil values as well as English, with the last 3 years preparing them for proper employment. Weekend and summer workshops are for acting, singing, dancing and painting - personality building is essential for children who live in depressed circumstances. Officers and family of globalization industry participate in the coaching[21]. The possibilities are clear to see in the Akanksha Centers, how the children know where they come from and where they are going: UP. In India, the role model is Abdul Kalam, who came from modest conditions, became a space engineer, and is now President of the more than a billion people of India[22]. We have an Up with Children organization to spread the Akanksha idea to other countries, including the U.S., but have not as yet had much success in fundraising and in finding sponsoring organizations in any cities[23].

Can we afford to eliminate the chasm between poor and rich? A solution to a problem will have to be fitting to the amount of money that is available to reach its goals, while the numbers of people all over the world needing help is enormous. The comparison of daily incomes in Table 1.1 indicates that the worst global poverty would disappear if the wealthiest billion people would contribute 10% of their income. The conclusion is therefore that the worst poverty in the world could be fixed quickly in principle, that we could afford to accomplish that goal. In practice, however, it is out of the question that enough of such large donors could be found. **Professional soliciting** and campaigning are essential to obtain a substantial number of large donations, or a large number of small contributions. Here comes the surprise, for the most interesting and efficient way to pay for it all, in the next section.

16.8 We can Win

So here comes our ultimate celebration like Beethoven's, but now regarding our world, our global environment. Nearly 121,000 words were used, that was a lot of theory and now it is time for some practical **doing.** Most of the solutions are costly – how are we going to pay for them? I propose a simple solution for that, but having great benefits and broad side effects. This is to levy a **100.0% tax on the sale of all weaponry and munitions or parts or facilities for same,** large or small, anywhere and at each time they are sold, without exemptions, to be taxed locally and the proceeds to be used locally against poverty or global warming. This should be watched against dodging, it should have a clause for judgment towards the spirit in which it is intended. The tax collects from the U.S. and its allies alone would presently amount to some $800 billion per year, on the assumption of selling twice on average, yielding $2 per day for special programs such as to coaching slum children towards leadership.

In case you think I am naïve, consider Sec. 15.3 again, and October 28, 1962 in Sec. 16.3, that we must disarm or perish, and that time is running out. Politically this would be **an attractive proposal** because everyone would see the logic of the tax and its application; votes in favor of this tax will be easy to obtain. It decreases killing and violence and it diminishes poverty and global warming, without lowering the standard of living; it is a winning proposition. If there were ever a problem with voting or action for the plan, ask the mothers to decide. It would be a detriment to senseless killing at high schools and colleges. The employment of military personnel could still be used in projects for coaching the children and working on humanitarian projects. For the United States, it would be a fulfillment of what its Founding Fathers and President Eisenhower dreamed.

The results would give a clear signal of the direction in which humanity is moving, from **profits gained in arms trade** to **progress in peace and prosperity**. Imagine the worldwide reactions! Terrorism would diminish because it would appear in a different light for those who might otherwise contemplate participation. An impression of civilized respect for the human enterprise will be the result of helping others attain what we wish for ourselves.

This book has been laying out problems and wishes for the world, or someone, to fix for us - but reciprocity suggests that we **begin with ourselves**. It is not just that then things might happen that we want - they will not happen unless we begin with ourselves. The global example is of course the U.S. - reciprocity suggests that it makes changes first. The voters know what to do; information and guidelines are in several recent books and in this one.

And so we end this book with some hope and cheer, and with our personal dedication to improving the future. Just because the problems seem to be overwhelming, we must dedicate ourselves such that we are ready for the proper action and get in the habit of looking out for things to do, and do them. We must never give in to feelings of hopelessness and helplessness; a bad time is a good time to prepare for better times. The strong and thoughtful will find solutions and make resources. The diligent often seem rewarded with some **good luck** as well. I am referring here to the attribute of nature that it is unpredictable, always in trial-and-error, and sometimes jumping forward, likely in the direction into which the people are working. To live the ethic in daily life becomes a pervasive hobby, a habit towards healthier living and participation in actions. Even simple ones are useful, like those of conservation and recycling, while it brings in return a positive outlook with a reasonable faith in the future[24].

The ethic is expressed with deeper sophistication but easy to remember in three words spoken by Mori Daiko, High Priest at the Shinsho-ji Temple in the countryside of Hiroshima. Once upon a time, we were welcome inside that expansive temple for an astronomical conference, and I was granted a personal interview with him. At the end of our meeting, he answered my respectful "Good Bye" with saying deliberately "Be the Sun." I knew that Buddhism and Hinduism are Sun-oriented, but for an astronomer it is a challenge to be the Sun. I asked Swami Manuvarya, who was dedicated to helping and advising hundreds of people of all ranks and education in his Ashram in Ahmedabad, India, if he had ever tried to be the Sun. He laughed it off, pointing at his rather worn garb and said, "oh no, for I am just a simple swami." However, he expressed it with such radiance and joyful security, that to me he *was* the Sun.

Be the Sun

REFERENCES AND NOTES

INTRODUCTION

1. Sarabhai, V.A., 1974. *Science Policy and National Development.* (Delhi: MacMillan), which is based on his earlier writing and actions. For Kapitza, P.L., see his obituary by Rotblat, J., in *Phys. Today* **37**, 95, 1984. Weisskopf, V.F., 2003. The Privilege of being a Physicist. *Phys. Today* **56, No. 2,** 48-52.

2. Gehrels, T., 1988. *On the Glassy Sea, an astronomer's journey* (New York: American Institute of Physics), ISBN 0-88318-598-9.

3. Section 8.4 explains how laws generally are summaries of observations. The natural laws are discussed in Chaisson, E., 2001. *Cosmic Evolution. The Rise of Complexity in Nature* (Boston: Harvard Univ. Press). An update is in Chaisson, E., 2006. *Epic of Evolution. Seven Ages of the Cosmos.* (New York: Columbia Univ. Press).

4. Texts at the most advanced level written by the scientists themselves are in the Space Science Series (Tucson: Univ. Ariz. Press); I will refer in these Notes to its recent volumes. Netscape and Internet are useful only if the source is clear, which should be as much as possible from the original discoverer or analyst. I recommend Shirley, J.H. & Fairbridge, R.W., eds., 1997. *Encyclopedia of Planetary Sciences.* (London: Chapman & Hall), and Weissman, P.R, McFadden, L.-A. & Johnson, T.V., eds., 1999. *Encyclopedia of the Solar System.* (New York: Academic Press). For life forms there are good texts such as of Purves, W.K., Sadava, D., Orians, G.H. & Heller, H.C., 2004. *Life: The Science of Biology,* 7th Ed. (Salt Lake City, UT: W.H. Freeman Co.). Keeping up with new ideas in many of our topics is easily done by reading *Scientific American,* which is available also on <www.sciam.com>,<www.sciamdigital.com>, <www.sciammind.com>; <http://antwrp.gsfc.nasa.gov/apod/> presents the "Astronomy Picture of the Day".

CHAPTER 1

1. Recent information and a haunting illustration regarding the earliest of all emigrants, the people having left Africa, are in Wong, K., 2003. Stranger in a New Land. *Sci. American* **289, No. 5**, 74-83.
2. Renner, M. 1997. *Fighting for Survival: Environmental Decline, Social Conflict, and the New Age of Insecurity.* (New York: W.W. Norton & Co.).
3. People who (think that they) have problems with math may find help in Mazur, B., 2003. *Imagining Numbers.* (New York: Farrar, Straus and Giraux). The convention for exponent is that they are written in terms of **a number between 1 and 10 multiplied by a power of 10**; it is a simple clean up, for instance the number 130 is 1.3 x 10^2, not 13 x 10. The Table gives an overview of exponents. The name "nano" is used for wavelengths when they are given in nanometers, nm, which is the case for the name "molecular nanotechnology" discussed in Sec. 14.5.

Quite helpful is to watch the filmstrip **"Powers of Ten"**, which begins with people, showing larger sizes out to the universe, and comes back showing sizes down to the atoms within people; it is available in libraries or Interlibrary Loan, or on Websites that indicate where it can be bought.

Table. Powers of Ten

10^0 = one = 1
10^1 = ten = 10
10^{-1} = tenth = 1/10
10^2 = hundred = 100
10^{-2} = hundredth = 1/100
10^3 = thousand = 1,000
10^{-3} = thousandth = 1/1,000
10^6 = million = 1,000,000
10^{-6} = millionth = 1/1,000,000
10^9 = billion = 1,000,000,000
10^{-9} = billionth = 1/1,000,000,000, "nano"
10^{12} = trillion = 1,000,000,000,000

4. Lunine, J.I., 1999. *EARTH, Evolution of a Habitable World* (Cambridge, U.K., Cambridge Univ. Press).
5. We use grams and centimeters, and rarely kilograms and meters, because they are not often needed in this book and then mostly for densities, in grams per cubic centimeter, with 1 gm/cm^3 for water and 0.99 gm/cm^3 for people which are easiest remembered and associated with. For temperatures, we

will have to go beyond the range we normally experience on the earth and so we need the Kelvin scale, named after Lord Kelvin (1824-1907) because it is based on **Absolute Zero**, an interesting concept because it is the temperature at which normal activity stops. It is therefore difficult if not impossible to reach and to measure; there are special laboratories where the physics near 0 K is studied and presently reached to within a fraction of a degree, about 0.03 K (no degree sign is used for Kelvin). Temperature conversions are easiest remembered with the Figure.

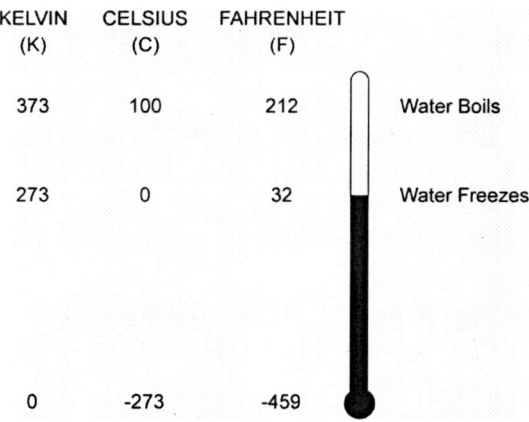

Anders Celsius (1701-1744) set the temperature of melting ice at zero and of boiling water at 100; his degrees are referred to as **Celsius**, but also sometimes as **centigrade** because of that number 100. These two, 0°C and 100°C happen to be 32°F and 212°F on the scale of Gabriel Fahrenheit (1686-1736), who had set 0 and 100 at what was then known for the coldest possible temperature and for the temperature of healthy human beings. That difference in Fahrenheit happens to be exactly 180° while the Celsius difference between the two points is 100 degrees. It is then noted that they both can be divided by 20 so that the scales differ as 9 to 5; Fahrenheit degrees are almost half as small as in centigrade. And we still have to take care of the 32 degrees difference as is shown in the Figure. For example, to convert 86°F to C, first subtract 32 to get 54, then divide by 9 to get 6 and multiply by 5, which gives us 30 degrees Celsius. And a friend in Europe writes it's a nice day, 10°C; divide that by 5 to get 2, multiply by 9 to get 18 and add 32 for a total of 50°F, chilly for Tucson. For the approximately 740 K for Mercury and Venus, subtract 273 to get 467°C, divide by 5 for 93.4, multiply by 9 to give 841 and add 32 for 873°F, which is much hotter than in our kitchen ovens.

6. The magnetic field of the earth is shown beautifully in Glatzmaier, G.A. & Olson, P., 2005. Probing the Geodynamo *Sci. American* **292, No. 4**, 50-57. Jokipii, J.R., Sonett, C.P., & Giampapa, M.S. eds., 1997. *Cosmic Winds and the Heliosphere* (Tucson: Univ. Ariz. Press). Bhatnagar, A. & Livingston, W.C., 2005. *Fundamentals of Solar Astronomy* (Hackensack, N.J.: World Scientific). The STEREO spacecraft are described in Morning, F, Jr., 2005. Solar Studies. *Aviation Week & Space Techn.* **163, No. 9**, 62-64. A spacecraft that is dedicated to solar-system wide effects, IMAGE, has as website <http://pluto.space.swri.edu/IMAGE/>.

7. Fascinating observations can be made when flying in airplanes. They are described in Wood, E.A., 1975. *Science from Your Airplane Window*, 2nd ed. (New York: Dover Publs.). Exceptional aurora pictures are shown in Williams, E. R., 2001. Sprites, Elves, and Glow Discharge Tubes, *Phys. Today* **54**, No. 11, 41-47. Lightning is explained in Dwyer, J.R., 2005. A Bolt out of the Blue *Sci. American* **292, No. 5**, 64-71.

8. Flessa, K.W. (pers. comm., 2001), who is a modern explorer of the estuary of the Colorado River.

9. *National Geographic Atlas of the World* (Washington, D.C.: Nat. Geogr. Soc.) has beautifully illustrated pages on Earth's crust and continents adrift. Also see Earle, S.A., 2001. *National Geographic Atlas of the Ocean, The Deep Frontier.* (Washington, D. C.: Nat. Geogr. Soc.). Carson, Rachel, 2003. *The Sea Around Us: An Illustrated Commemorative Edition* (New York: Oxford Univ. Press); this is an update of a 1951 book; its Afterword gives a lucid explanation of plate tectonics.

10. Gehrels, G.E., DeCelles, P.G., Martin, A., Ojha, T.P., Pinhassi, G., & Upreti, B.N., 2003, Initiation of the Himalayan orogen as an early Paleozoic thin-skinned thrust belt, *GSA Today* **13, No. 9**, 4-9.

11. <http://earthquake.usgs.gov/faq/meas.html>

12. Fabulous pictures of hydrothermal vents are shown in Lutz, R.A. *et al.*, 2003. Dawn in the Deep. *Nat. Geographic* **203, No. 2**, 92-103.

13. The website of the U.S. Geological Survey's Hawiian Volcano Observatory is
<http://wwwhvo.wr.usgs.gov/volcanoes/loihi/main.html>.

14. Gehrels, T. & Coffeen, D.L., 1969. Ultraviolet polarimetry of planets. *COSPAR Space Res. X*, 1036-1042.

15. Kroodsma, D.E., 2005. *The singing life of birds: the art and science of listening to birdsong* (Houghton Miflin); one can listen with its compact disc.

CHAPTER 2

1. Brownlee, D., 1985. Cosmic Dust: Collection and Research, *Ann. Rev. Earth Planet. Sci.* **13**, 147-173.
2. Lauretta, D. & McSween Jr., H.Y., eds. 2006. *Meteorites and the Early Solar System II* (Tucson: Univ. Ariz. Press).
3. The meaning of "law" is explained in Sec. 8.4. A suspicion that Kepler poisoned Brahe is brought forward by Gilder, J. & Gilder A.-L., 2004. *Heavenly Intrigue: Johannes Kepler, Tycho Brahe, and the Murder behind One of History's Greatest Scientific Discoveries* (New York: Doubleday). Also see Maldacena, J., 2005. The Illusion of Gravity. *Sci. American* **293, No. 5**, 56-63.
4. Soter, S., 2007. What is a Planet? *Sci. American* **296, No. 1**, 34-41. Gehrels, T., 2005. Dual role for Pluto in the great planetary debate. *Nature* **436, No. 7054**, 1088. Gehrels, T., 2006. What is a Planet? *Sky & Tel.* **111, No. 1**, 14. Gehrels, T., 2006. Pluto the Ninth, Xena (2003 UB_{313}) the Tenth, and brighter than Pluto after that. *Dissertatio cum Nuncio Sidereo III*, official newspaper of the IAU 2006.
5. Whitaker, E. A., 1999, *Mapping and naming the Moon* (Cambridge Univ. Press).
6. Whitaker, E. A., pers. comm. 2001.
7. Vilas, F., Chapman, C.R. & Matthews, M.S., eds., 1988. *Mercury* (Tucson: Univ. Ariz. Press). Miller, J., 2007. Radar reveals Mercury's molten core. *Physics Today* **60, No. 7**, 22-24.
8. <www.nasa.gov/mission_pages/messenger/main/index.html>
<http://sci.esa.int/science-e/www/area/index.cfm?fareaid=30>
9. Bougher, S.W., Hunten, D.M. & Phillips, R.J., eds., 1997. *Venus II - Geology, Geophysics, Atmosphere, and Solar Wind Environment* (Tucson: Univ. Ariz. Press).
10. Kieffer, H.H., Jakosky, B.M., Snyder, C.W. & Matthews, M.S., eds., 1992. *Mars* (Tucson: Univ. Ariz. Press). Also see Christensen, P.R., 2005. The Many Faces of Mars. *Sci. American* **293, No. 1**, 32-39, Naeye, R., 2005. Europe's Eye on Mars *Sky & Tel.* **110, No. 6**, 31-36, and Covault, C., 2005. Spirits Soar, *Aviation Week & Space Techn.* **163, No. 19**, 48-53. Bell, J., 2006. The Red Planet's Watery Past. *Sci. American* **295, No. 6**, 62-69.
11. Hodges, K. 2006. Climate and the evolution of Mountains. *Sci. American* **295, No.2**, 72-79.

CHAPTER 3

1. Potter, A., & Morgan, T., 1985. Discovery of sodium in the atmosphere of Mercury *Science* **229**, 651-653. Potter, A. & Morgan, T., 1988.

Discovery of sodium and potassium vapor in the atmosphere of the moon *Science* **241**, 675-680, with a unique image. Butler, B., 2004. Ice on Mercury and the Moon In Dasch, P., ed., 2004. *Icy worlds of the solar system* (Cambridge, U.K.: Cambridge Univ. Press).
2. Atreya, S. K, Pollack, J. B. & Matthews, M. S., eds., 1989. *Origin and Evolution of Planetary and Satellite Atmospheres* (Tucson: Univ. Ariz. Press). Atreya, S.K., 2007. The Mystery of Methane on Mars and Titan, *Sci. American* **296, No. 5**, 42-51.
3. Carr, M. H., 1979. Formation of Martian flood features by release of water from confined aquifer, *J. Geophys. Res.*, **84**, 2995-3007.
4. Baker *et al.*, 1991. Ancient oceans, ice sheets and the hydrological cycle on Mars, *Nature*, **352**, 589-594.
5. See Note 14 of Ch. 1.
6. <www.esa.int/SPECIALS/Venus_Express>
7. See ref. 4 of Ch. 1, and Ksanfomality, L.V., 1997. Possible Emergence and Evolution of Life are restricted by Characteristics of the Planet, *Astrophys. & Space. Sc.* **252**, 41-50.

CHAPTER 4

1. Cunningham, C.J., 2001. *The First Asteroid.* (Surside, Florida: Star Lab Press).
2. Bottke W.F., Cellino, A., Paolicchi, P. & Binzel, R.P., eds., 2002. *Asteroids III* (Tucson: Univ. Ariz. Press).
3. The name "asteroiders" was initiated in the Foreword of the previous Ref.
4. Merline, W.J., *et al.*, 2002. Asteroids Do Have Satellites, in the book of the above Note 2, pp. 289-312. Rubin, A.E., 2005. What Heated the Asteroids? *Sci. American* **292, No. 5**, 80-87. Covault, C., 2007. Blue Light Special. *Aviation Week & Space Techn.* **167. No.** 1, 56-61; the title is because ion propulsion is used for the DAWN spacecraft.
5. Bagenal, F., McKinnon, W. & Dowling, T., eds., 2002, *Jupiter: Planet, Satellites and Magnetosphere* (Cambridge Univ. Press.). Information about that book is in <http://lasp.colorado.edu/JUPITER/index.html>.
6. Gehrels, T. & Matthews, M.S., eds., 1984. *Saturn* (Tucson: Univ. Ariz. Press). Updates are in Shirley, J.H. & Fairbridge, R.W., eds., 1997. *Encyclopedia of Planetary Sciences* (London: Chapman & Hall), and in Weissman, P.R, McFadden, L.-A. & Johnson, T.V., eds., 1999. *Encyclopedia of the Solar System.* (New York: Academic Press).
7. Bergstralh, J.T., Miner, E.D. & Matthews, M.S., eds., 1991. *Uranus,* (Tucson: Univ. Ariz. Press).
8. Cruikshank, D.P., ed., 1995. *Neptune and Triton* (Tucson: Univ. Ariz.

Press).

9. Burns, J.A. & Matthews, M.S., eds., 1986. *Satellites* (Tucson: Univ. Ariz. Press). For an update see the *Encyclopedias* in Note 6. Jewitt, D., Sheppard, S.S. & Kleyna, J. The strangest Satellites in the Solar System. *Sci. American* **295, No. 2**, 41-47. <www.igpp.ucla.edu/europa>. Pappalardo, R., McKinnon, W. & Khurana, K., eds., 2009. Europa (Tucson: Univ. Ariz. Press).

10. Greenberg, R. & Brahic, A., eds., 1984. *Planetary Rings* (Tucson: Univ. Ariz. Press), and see Note 6.

11. Marsden, B.G. (pers. comm., 2007).

12. Kowal, C.T., 1996. *Asteroids, Their Nature and Utilization,* 2nd ed., (New York: John Wiley & Sons); Kowal is the discoverer of Chiron.

13. <www.boulder.swri.edu/ekonews/issues/past/n010/html/index.html> Kuiper, G.P., 1951. On the Origin of the Solar System. In J.A. Hynek, ed. *Astrophysics* (New York: McGraw-Hill Book Co.), 367-424. Barucci, M.A., Boehnhardt, H., Cruikshank, D. & Morbidelli, A., eds. 2008, The Solar System beyond Neptune (Tucson: Univ. Ariz. Press); see <www.lesia.obspm.fr/planeto/KuiperBook/index.html>.

14. Stern, S.A. & Tholen, D.J., eds., 1997. *Pluto and Charon* (Tucson: Univ. Ariz. Press).

15. Smoluchowski, R., Bahcall, J.N. & Matthews, M.S., eds., 1986. *The Galaxy and the Solar System* (Tucson: Univ. Ariz. Press).

16. Festou, M., Keller, H.U. & Weaver, H.A., eds., 2005. *Comets II* (Tucson: Univ. Ariz. Press).

17. His original work was updated in Oort. J.H., 1963. Empirical Data on the Origin of Comets, in *The Moon, Meteorites and Comets,* Middlehurst, B.M. & Kuiper, G.P. eds. (Univ. Chicago Press), 665-673.

18. The maneuvering and images obtained by the Deep Impact spacecraft are in Dornheim, M.A., 2005. Crash Course *Aviation Week & Space Techn.* **163, No. 2,** 28-31.

CHAPTER 5

1. Cox, A.N., Livingston, W.C. & Matthews, M.S., eds., 1991. *Solar Interior and Atmosphere* (Tucson: Univ. Ariz. Press).

2. Sonett, C.P., Giampapa, M.S. & Matthews, M.S., eds., 1991. *The Sun in Time* (Tucson: Univ. Ariz. Press).

3. Parker, E.N., 2000. The Physics of the Sun and the Gateway to the Stars, *Phys. Today*, June, 26-31.

4. See <www.solarphysics.kva.se>, and for Procyon and the white dwarf see <www.eso.org/outreach/eduoff/catchastar/cas-projects/uk_procyon_1/>.

5. Hillebrandt, W., Janka, H.-T. & Müller, E., 2006. How to Blow Up a Star. *Sci. American* **295, No. 1**, 42-51, presents various models of supernovae.
6. Blandford, R. & Gehrels, N. 1999. Revisiting the Black Hole. *Phys. Today* **52, No. 6**, 40-46. Gehrels, N., Piro, L. & Leonard, P.J.T., 2002. The Brightest Explosions in the Universe. Every time a gamma-ray burst goes off, a black hole is born. *Sc. Am.* **287, No, 6**, 84-91. Carr, B.J. & Giddings, S.B., 2005. Quantum Black Holes. *Sci. American* **292, No. 5**, 48-55; small black holes may be produced in the laboratory.

CHAPTER 6

1. Kawaguchi, K. *et al.*, 1993. Detection of a new molecular ion HC_3NH^+ in TMC-1, *Astrophys. J.,* **420**, L95-97.
2. The distance of 26,080 is adopted by Cox, A. N., ed., 2000. *Allen's Astrophysical Quantities*, 4th ed. (New York: Springer-Verlag), p. 557. Spiral structure is explained in Combes, F., 2005. Ripples in a Galactic Pond. *Sci. American* **293, No. 4,** 42-49.
3. See Hawking, S, 1988. *A Brief History of Time. From the Big Bang to Black Holes* (New York: Bantam Books), and Kochanek, C. S., & Hewitt, J. N., 1995. *Astrophysical Applications of Gravitational Lensing.* (Dordrecht, Netherlands: Kluwer Academic).
4. National Geographic Atlas of the World, 6th revised edition, 1992. (Washington, D.C.: Nat. Geog. Soc.) p.119. Distributions of galaxies is also shown by the Sloan Survey at Apache Point, New Mexico, in Strauss, M.A., 2004. Reading the Blueprints of Creation. *Sci. American* **290, No. 2**, 54-61.
5. Schweizer, F., 2000. Interactions as a driver of galaxy evolution, *Phil. Trans. R. Soc. London, A* **358,** 2063-2076.
6. Gehrels, N. & Paul. J., 1998. The New Gamma-Ray Astronomy. *Phys. Today* **51,** 26-32. NASA's Explorer missions are for astrophysics and space physics, with MIDEX, cost cap about $220M, presently Chandra, Swift, Spitzer, and GALEX (Galaxy Evolution Explorer); SMEX, ~$150M beginning in 2008; Discovery missions are for solar system exploration, NASA has Discovery missions ~$425M, New Frontiers missions, ~$700M, with New Horizons to Pluto as the first one.
7. <http://library.thinkquest.org/C003763/index.php?page=planet08> shows animation and wavelength changes between red shift and blueshift. The acceleration in the expansion of extragalactic space is described in Riess, A.G. & Turner, M.S., 2004. From Slowdown to Speedup. *Sci. American* **290, No. 2,** 62-67. The explanation by leakage of gravity is in Dvali, G., 2004. Out of the Darkness. *Sci. American* **290, No. 2**, 68-75. Conselice,

C.J., 2007. The Universe's Invisible Hand. *Sci. American* **296, No. 2**, 34-41, the cosmic grip of dark energy even on galaxies.

8. A general training is obtained from Narlikar, J.V., 2002. *An Introduction to Cosmology*, 3rd ed. (Cambridge Univ. Press). The Big Bang cosmology is described in several books, such as M. Rees, M., 1997. *Before the Beginning* (Reading, Mass.: Addison-Wesley). Continuous creation is in Burbidge, G., Hoyle, F, Narlikar, J.V. , 2000. *A Different Approach to Cosmology, from a Static Universe through the Big Bang towards Reality* (Cambridge Univ. Press); there had been critique by a variety of experts, in "Letters" of the *Physics Today* issues for April and September 1999.

9. Smolin, L., 1997. *The Life of the Cosmos* (New York: Oxford Univ. Press).

10. Aguirre, A. & Gratton, S., 2002. Steady-State Eternal Inflation. *Phys. Rev.* **D 65,** also: Inflation without a beginning: a null boundary proposal. *Phys. Rev.* D **67**, 083515 <gr-qc/0301042> 2003. Guth, A.H., 1997. *The inflationary universe: the quest for a new theory of cosmic origins* (New York: Addison-Wesley). Guth, A.H., in Measuring and Modeling the Universe, ed. W.L. Freedman (Cambridge Univ. Press, Cambridge, U.K., 2003), astro-ph/0405546. Linde, A. 2007, book review of Vilenkin, A., 2006. *Many Worlds in One, the Search for Other Universes.* (New York: Hill and Wang).

11. Gehrels, T., 2007. Universes seen by a Chandrasekhar equation in stellar physics. <http://arXiv.org/astro-ph/0701344>. Also Gehrels, T., 2007. The Multiverse and the Origin of our Universe. http://arXiv.org/abs/0707.1030.

CHAPTER 7

1. Reference 3 of the Introduction has the Chaisson references.

2. Wallace, A.R., 1973. Contributions to the theory of natural selection; a series of essays. (New York: AMS Press), originally 1870. Kutschera, U., 2003. A Comparative Analysis of the Darwin-Wallace Papers and the Development of the Concepts of Natural Selection. *Theory Biosc.***122**, 343-359. Also see <www.wku.edu/~smithch/index1.htm> for The Alfred Russel Wallace Page.

3. Christian, D., 2004, *Maps of Time. An Introduction to Big History.* (Univ. California Press). Spier, F., 2005. Energy Flows and the Rise and Demise of Complexity. *Social Evolution & History* **1, No. 1**.

4. Smuts, J.C., 1973. *Holism and Evolution* (Westport, Connecticut: Greenwood), originally published in 1926.

5. Kauffman, S.A., 1993. *The Origins of Order, Self-Organization and Selection in Evolution..* (Oxford, U.K.: Oxford Univ. Press). Also by

Kauffman at Oxford Univ. Press, 1995, *At Home in the Universe, The Search for Laws of Self-Organization and Complexity.*
6. Narlikar, J.V. & Apparao, K.M.V., 1982, studied the transition from a contracting to an expanding universe in *Astrophys. Space Sc.* **87**, 333.
7. Greene, B., 1999. *The Elegant Universe, Superstrings, Hidden Dimensions, and the Quest for the Ultimate Theory.* (New York: Random House). A delightful Web site is <www.superstringtheory.com>.
8. Gell-Mann, M., 1994. *The Quark and the Jaguar* (New York: W.H. Freeman & Co.).
9. Randall, L., 2005. *Warped Passages* (New York: Harper-Collins). Also see Randall, L. The case for extra dimensions. *Physics Today* **60**, No. 7, 80-81. Steinhardt, P.J. & Turok, N., 2007. Endless Universe, Beyond the Big Bang. (New York: Doubleday). However, also see Woit, P., 2006. Not even wrong, the failure of string theory and the search for unity in physical law. (New York: Basic Books).
10. See Note 3 of Ch. 6.
11. Schwarzschild, B., 2003. WMAP Spacecraft Maps the Entire Cosmic Microwave Sky With Unprecedented Precision. *Physics Today* **56**, No. 4, 21-24.
12. <www.mpa-garching.mpg.de>

CHAPTER 8

1. The first suggestion that matter is composed of waves is in Clifford, W., 1870. On the Space Theory of Matter. *Cambridge Philosoph. Soc.*, **2**, 157-158. Outward and inward waves are discussed in Wheeler, J. A. & Feynman, R. P., 1945. Interaction with the Absorber as the Mechanism of Radiation, *Rev. Mod. Phys.* **17**, 157-181, in Cramer, J., 1986. Transactional interpretation of quantum mechanics. *Rev. Mod. Phys.* **58**, 647-687, and in Wolff, M., 1995. Beyond the Point Particle – A Wave Structure for the Electron. *Galilean Electrodynamics* **6, No. 5,** 83-91.
2. Ellen Williams, personal communication, 1984.
3. <http://map.gsfc.nasa.gov/> Also see Perlmutter, S., 2003. Supernovae, Dark Energy, and the Accelerating Universe. *Phys. Today* **56**, No. 4, 53-60. A good overview of the composition of the universe is in Cline, D.B., 2003. The search for Dark Matter *Sci. American* **288**, **No. 3**, 50-59. Conselice, C.J., 2007. The Universe's Invisible Hand. *Sci. American* **296, No. 2**, 34-41, the cosmic grip of dark energy even on galaxies.
4. Fisher, H., 2004. *The Brain in Love* (New York: Henry Holt). Regarding discussions of Bohr and Heisenberg, M. Fayn wrote the play *Copenhagen*, which is also available in VHS version from PBS Hollywood.

5. The explanation that light should be understood as waves of radiation rather than as photon particles in Lamb, W.E., Jr., 1994. Anti-photon, *Appl. Phys. B* **60**, 77-84.

6. A collection of references to original papers is in Bell, J.S., 1987. *Speakable and Unspeakable in Quantum Mechanics* (Cambridge University Press). For a modern treatment that supercedes some of the classical studies see Lamb, W.E., Jr., 1996. Classical Theory of Measurement: A big step towards the Quantum Theory of Measurement, in Chiao, R.Y., ed., 1996. *Amazing Light* (New York: Springer-Verlag).

7. Feynman, R., 1967. *The Character of Physical Law.* (Cambridge, MA: M.I.T. Press), p.129.

8. The quote of Max Planck is in Gross, D.J., 1989. Can we scale the Planck scale? *Phys. Today* **42**, **No. 6**, 9-11. The original was in Planck, M., 1899. Über irreversibele Strahlungsvorgänge, *Sitzungsber. Preusz. Akad. Wissenschaften*, **5**, 479.

9. Kane, G., 2005. The Mysteries of Mass *Sci. American* **293, No. 1**, 41-48.

10. See Note 3 of Ch. 2 for the suspicion that Kepler poisoned Brahe.

11. Mach, E., 1983. *The Science of Mechanics.* (Chicago: Open Court Publ.); original in German, 1883.

12. Hu, W. & White, M., 2004. The Cosmic Symphony. *Sci American* **290, No. 2**, 44-53.

The appeal and power of music are explained in Weinberger, N.M., 2004. Music and the Brain. *Sci. American* **291, No. 5**, 88-95. Also see Note 15 of Ch. 1.

CHAPTER 9

1. Zepf, S.E. & Ashman, K.M., 2003. The Unexpected Youth of Globular Clusters. *Sc. American* **289, No. 4**, 46-51, beautifully illustrated.

2. Ref. 5 of Ch. 6.

3. Reipurth, B., Jewitt, D. & Keil, K., eds., 2006. *Protostars and Planets V* (Tucson: Univ. Ariz. Press). Bally, J. & Reipurth, B., 2007. *The Birth of Stars and Planets* (Cambridge Univ. Press).

4. Gehrels, T., ed., 1974. *Planets, Stars and Nebulae, Studied with Photopolarimetry* (Tucson: Univ. Ariz. Press).

CHAPTER 10

1. Sonett, C.P., M. S. Giampapa, M.S. & Matthews, M.S., eds. 1991. *The Sun in Time* (Tucson: Univ. of Ariz. Press).

2. Ref. 2 of Ch. 2.

3. Levinson, H. F., *et al.*, 2001. Could the lunar 'Late heavy Bombardment'

have been triggered by the Formation of Uranus and Neptune? *Icarus*, **151**. 286-306.

4. Canup, R.M. & Righter, K., eds. 2000. *Origin of the earth and moon.* (Tucson: Univ. Ariz. Press).

5. A thorough description is in Lunine, J.I., 1999. *EARTH. Evolution of a Habitable World.* (Cambridge, U.K.: Cambridge Univ. Press). Also see Valley J.W., 2005. A Cool Early Earth? *Sci. American* **293, No. 4**, 58-65. Maps of the North and South America plates 200, 140, and 80 My ago are in Van Arsdale, R.B. & Cox, R.T., 2007. The Mississippi's Curious Origin. *Sc. American* **296, No.1**, 76-82B.

CHAPTER 11

1. Gold, T., 1999. *The Deep Hot Biosphere.* (New York: Springer-Verlag). Fascinating photos of hydrothermal vents and fish in the deep ocean can be seen in Lutz, R.A., 2003. Dawn in the Deep. *Nat. Geographic* **203, No. 2**, 92-103. Kasting, J.F., 2004. When Methane Made Climate. *Sci. American* **292, No. 1**, 78-85.

2. Delsemme, A.H., 1998. *Our cosmic origin: from the Big Bang to the emergence of life and intelligence* (Cambridge, UK: Cambridge Univ. Press). Also see Delsemme, A.H., 1999. Cometary origin of the biosphere. *Icarus* 146, 313-325.

3. Lovelock, J. E., 1990. Hands up for the Gaia hypothesis, *Nature* **344**, 100-102.

4. Darwin, C., 1839. *The Voyage of the Beagle.* (I have used a 1989 print, London: Penguin Books). An interesting overview of the culture at the time is reprinted in 1994. Chambers, R., 1994. *Vestiges of the Natural History of Creation and other Evolutionary Writing* (Chicago: Univ. Chicago Press).

5. Darwin, C., 1962. *The Origin of Species by Means of Natural Selection, or the Preservation of Favoured Races in the Struggle for Life.* (London: Collier MacMillan is the edition I used).

6. The version that has the removed sentences restored is by his granddaughter Nora Barlow, 1958. *The autobiography of Charles Darwin* (New York: W.W. Norton).

7. The Web site for the Institute for Creation Research is <www.icr.org>. The National Center for Science Education presents the opposite view in <www.ncseweb.org>.

8. See Note 2 of Ch. 7.

9. Purves, W.K. *et al.*, 2004. *LIFE, the Science of Biology,* 7th Edition (Gordonsville, Virginia: W.H. Freeman & Co.), Watson, J.D., 1976, *Molecular Biology of the Gene* (Menlo Park, Calif.: W.A.Benjamin), and

Watson and Crick describe structure of DNA, 1953, in <www.pbs.org/wgbh/aso/databank/entries/do53dn.html>.

10. Kutter, G.S., 1987. *The Universe and Life* (Boston: Jones and Bartlett Publ.).

11. Cahill, L., 2005. His Brain, Her Brain *Sci. American* **292, No. 5**, 40-47. Kinsley, C.H., & Lambert, K.G., 2006. The Maternal Brain. *Sci. American* **294, No.1**, 72-79. Ross, P.E., 2006. The Expert Mind. (Experts are made, not born.) *Sci. American* **295, No. 2**, 64-71,

12. Lemonick, M. D. & Dorfman, A., 2001. One Giant Step for Mankind *Time* **158**, No. 3, 54-61. Wong, K., 2006. Lucy's Baby *Sci. American* **295, No.6**, 78-85 raises questions about the development of the time when upright walk occurred.

13. Mellars, P., 1998. The fate of the Neanderthals. *Nature* **395**, 539.

14. A haunting picture of an explorer 1.75 million years ago is shown in Wong, K., 2003. Stranger in a New Land. *Sc. American* **289, No. 5**, 74-83.

15. Spier, F., 1996. *The Structure of Big History: From the Big Bang until Today* (Chicago: Univ. Chicago Press).

16. de Waal, F.B.M., 2005. How animals do business. *Sci. American* **292, No. 4**, 72-79.

17. Bostrom, N., 2005. *The Future of Humankind* (New York: Doubleday).

18. Horgan, J., 2005. The Forgotten Era of Brain Chips. *Sci. American* **293, No. 4**, 66-73.

CHAPTER 12

1. Schweizer, F., 1998, in *Galaxies: Interactions and Induced Star Formation*, Friedli *et al.*, eds. (Berlin: Springer), pp. 105-274. Also see Schweizer, F., 2000. Interactions as a driver of galaxy evolution. *Phil. Trans. R. Soc. London A*, **358**, 2063-2076.

2. Gehrels, N. *et al.*, 2003. Ozone Depletion from Nearby Supernovae. *Astrophys. J.* **585**, 1169-1176.

3. Thomas. B.C. *et al.*, 2005. Gamma-Ray Bursts and the Earth: Exploration of Atmospheric, Biological, Climactic and Biogeochemical Effects. *Astrophys. J*, **634**, 509-533.

4. Balick, B. & Frank, A., 2004. The Extraordinary Deaths of Ordinary Stars. *Sci. American* **291, No. 1**, 50-61. Also see Ref. 3 of Ch. 5.

5. Smoluchowski, R, Bahcall, J.N. & Matthews, M.S. eds., 1986. *The Galaxy and the Solar System* (Tucson: Univ. Ariz. Press) is dedicated largely to the possibility of *periodic* impacts by comets and asteroids. Also, however, Ward, P.D. 2006. Impact from the Deep. (cause of mass extinctions other than asteroids) *Sci. American* **295, No. 4**, 64-71.

6. Various aspects of the asteroid hazard, from detection to mitigation, are treated in Gehrels, T., ed., 1994. *Hazards due to Comets and Asteroids*, (Tucson: Univ. Ariz. Press). The general NASA Web site is <http://neo.jpl.nasa.gov>. Also see Kring, D.A. & Durda, D.D. 2003. The Day the World Burned *Sci. American* **289, No. 6**, 98-105. <www.b612foundation.org> is the Web site for more aggressive defense; the Foundation is named after the B612 asteroid in de Saint-Exupéry, A., 1945. *Le Petit Prince*. (Paris: Gallimard). A new method for avoiding an impact has been suggested by Schweickart, R.L., Lu, E.T., Hut, P., & Chapman, C.R., 2003. The Asteroid Tugboat. *Sci. American,* **289, No. 5,** 54-59.
7. Levy, D. H., Shoemaker, E. M. & Shoemaker, C. S., 1995. Comet Shoemaker – Levy meets Jupiter. *Sci American* **273, No. 2**, 84-91.
8. Swindel, G.W. Jr. & Jones, W.B., 1954. The Sylacauga, Talladega County, Alabama, Aerolite. *Meteoritics* **1**, 125-132. Also see Povenmire, H., 2003. Sylacauga, Alabama Meteorite Revisited. *Meteorite* **9, No. 2**, 26-28.

CHAPTER 13
1. <www.csi.ad.jp/ABOMB/>
2. *Encyclopedia Americana, International Edition*. Nuclear War (Danbury, Connecticut), **20**, 519-533, 1993.
3. Lee. B., 1995. *Marching Orders* (New York: Crown Publ.) tells of negotiations about a possible Soviet– Japan alliance, which may have influenced the decision to terminate the war by using the bomb.
4. Robert McNamara, who was U.S. Secretary of Defense at the time, commented during a public broadcasting of February 22, 2001 how close we had come to nuclear war on October 28, 1962. Also see McNamara, R., 2004, *The Fog of War,* available in DVD format.
5. Smith D., (2001). "The World at War January 1, 2001". *The Defense Monitor* **30, No. 1**, 1-15, available on the article is followed by "Organizing for Peace," 17-18; www.cdi.org/dm/2001/issue1/world.html.
6. Creating a World Free of Nuclear Weapons: Global Security, *Union of Concerned Scientists,* 24 Oct. 2001, <www.ucsusa.org/missiledefense/>. Levi, M., 2004. Nuclear Bunker Buster Bombs *Sci. American* **291, No. 2**, 66-73. Rhodes, R., 2005. Living with the Bomb. How Safe is the World from the Ultimate Weapon? *Nat. Geographic* **208, No.1**. 98-113.
7. The Union of Concerned Scientists is a nonprofit partnership of scientists and citizens combining scientific analysis, innovative policy development and effective citizen advocacy to achieve practical environmental solutions, see <www.ucsusa.org>. Much of the present information is taken from

issues of *The Defense Monitor*, published by the Center for Defense Information, which is associated with *The World Security Institute*; *Def. Mon.* **28, No. 3**, 1999, is a Special 21st Century Edition. For deterrence rockets still aimed at each other, see *Def. Mon.* **33. No. 5,** 2004. A clear overview of military rocketry is in Garwin, R.L., 2004. Holes in the Missile Shield. *Sci. American* **291, No. 5**, 70-79.

8. *The Defense Monitor,* **28, No. 7**, 1999. The safety of nuclear waste storage in Yucca Mountain is in Nadis, S., 2003. Man against a Mountain *Sci. American* **288, No. 3**, 48-49. Uranium ammunitions are in Airhart, M., 2003. Cleaning Up after War. *Sci. American* **289, No. 4**, 44-45.

9. Dickey, C. & Thomas, E. 2002. How Saddam happened. *Newsweek*, Sept. 23 issue, 35-41.

10. Lyons, G. Invisible Wars: The Pentagon plays with poison. *Harper's* Dec. 1980 issue.

11. Bio-Warfare called 'Weapon of Choice'. *Av. Week & Space Technology*, April 12, 1999. Miller, J., Engelberg, S. & Broad, W., 2001. *Germs: Biological Weapons and America's Secret War.* (New York: Simon & Schuster).

12. Eisenhower, D.D., 1969. *In Review* (Garden City, N.Y.: Doubleday & Co.), p. 234; http://coursesa.matrix.msu.edu/~hst306/documents/indust.html has the entire speech.. Also see Carroll, J., 2001. Eisenhower's farewell warning was meant for our time. *Boston Globe*, January 16, reproduced in <www.commondreams.org/views01/0116-01.htm>. Congress' operations are in <www.ciaonet.org/olj/wpj/wpj_spring01b.html>.

13. From *The Ecologist*, 1999

14. Phillips, K. 2002. *Wealth and democracy: a political history of the American rich.* (New York: Broadway Books). The quote is from Zakaria, F. 2003. The Arrogant Empire. *Newsweek,* March 24, 18-23.

15. Nehru, J. 1942. *Glimpses of World History.* (Bloomington, Indiana: Indiana Univ. Press, 1962). At the 2004 Conference of the Historical Society, Professor Graydon Tunstall, Jr. of the University of South Florida presented a lecture on how one fails to find the records for the origins of WWI because they had been destroyed.

16. Swanson, G.J., 2004. *America the Broke* (New York: Double Day).

17. Hellman, C. 2002. The Pentagon's Fiscal Year 2003 Budget Request: More of Everything. *The Defense Monitor* **31, No. 2**, 2-4.

18. Johnson, C. 1999. *Blowback. The Costs and consequences of American Empire.* (New York: Henry Holt & Co.). How industry groups sometimes manufacture uncertainty about scientific findings is in Michaels, D., 2005. Doubt is their Product. *Sci. American* **292, No. 6**, 96-101.

19. The reference for my own earlier observations is in Note 2 of the Introduction.
20. *Los Angeles Times*, February 14, 1998. Also see the end of Sec. 16.6.
21. Barnett, T.P.M., 2004. *The Pentagon's New Map. War and Peace in the Twenty-first Century.* (New York: G.P. Putnam's Sons).
22. Steel, R. 1967. *Pax Americana* (New York: Viking).
23. Wilson, E.D., 1992. *Nat. Geogr. Atlas.* (Washington, D.C.: Nat. Geogr. Soc.), p. 5. Lovins, A.B., 2005. More Profit with Less Carbon. *Sci. American* **293, No. 3**, 74-83.
24. Strada, G., 1996. The Horror of Landmines. *Sci. American* **274, No. 5**, 40-45. Doney, S. C., 2006. The Dangers of Ocean Acidification. *Sci. American* **294, No. 3**, 58-65. The high smoke stacks are reported by *In Brief,* tri-annual newsletter of the Sierra Club, March 1983. Pumping CO_2 underground is in Socolow, R.H., 2005. Can we bury Global Warming? *Sci. American* **293, No. 1**, 49-55.
25. Moore, T.G., 1998. *Climate of Fear: Why we shouldn't worry about Global Warming* (Washington, D.C: Cato Institute). Global warming data are discussed in <www.ucsusa.org/environment/skeptics.html> and by DeGaetano, A.T. & Allen, R.J, 2002. Trends in twentieth-century temperature extremes across the United States, *Journal of Climate,* **15**, 3188-3205. An overview of problems and solutions in the southern parts of the U.S. is in <www.ucsusa.org/gulf>. Also see Hansen, J., 2004. Defusing the Global Warming Time Bomb. *Sci. American* **290, No. 3**, 68-77. Keppler, F., & Rockmann, T., 2007. Methane, Plants and Climate Change. *Sci. American* **296, No. 2**, 52-57. A focus on global warming and endangered species is provided in Slobodkin, L.B., 2003. *A Citizen's Guide to Ecology.* (New York: Oxford Univ. Press).
26. Lomborg, B., 2001. *The Skeptical Environmentalist. Measuring the Real State of the World.* (Cambridge, U.K.: Cambridge Univ. Press).
27. The Melting of the World's Ice, *Worldwatch,* November-December, 2000. Doyle, R., 2005. Melting at the Top. Rapidly warming Antarctic will have global consequences. *Sci. American* **292, No. 2**, 31. Trenberth, K.E., 2007. Warmer Oceans, Stronger Hurricanes. *Sci. American* **297, No. 1**, 44-51. Collins, W., *et al.,* 2007. The Physical Science behind Climate Change. *Sci. American* **297, No. 2**, 64-73.
28. Alley, R.B., 2004. Abrupt Climate Change. *Sci. American* **291, No. 5**, 62-69. *TIME* magazine devoted its Cover Story of April 9, 2001 to global warming. On its page 88, a Letter to President Bush with a plea for action was signed by J. Carter, M. Gorbachev, J. Glenn, W. Cronkite, G. Soros, J. C. Venter, J. Goodall, E. D. Wilson, H. Ford & S. Hawking. Also see Gore,

A., 2006. *An inconvenient truth: the planetary emergency of global warming and what we can do about it.* (Emmaus, Pennsylvania: Rodale Press). Strom, R. 2007. HOT HOUSE. *Global Climate Change and the Human Condition.* (New York: Springer Verlag). Kaplan, K.H., 2007. UE hammers out pact to cut greenhouse gas emissions. *Phys. Today* **60, No. 5**, 26-28; www.physicstoday.org.

29. Perrett, B., 2007. GREEN SKIES. The prospect of global emissions trading may emerge as a solution to the problem of greenhouse gases produced by airliners. *Aviation Week* **166, No. 15**, 28-29. The Journal of the Natural Resources Defense Council (NRDC), "This Green Life" may be surfed on <www.nrdc.org/thisgreenlife/>.

30. Global warming and endangered species are overviewed in Slobodkin, L.B., 2003. *A Citizen's Guide to Ecology.* (New York: Oxford Univ. Press). Pauly, D. & Watson, R. 2003. Counting the Last Fish. *Sci. American* **289, No. 1**, 42-47. Robbins, J., 2004. Lessons from the Wolf. *Sci. American* **290, No. 6**, 76-83 describes the effects on flora and fauna at Yellowstone from restoring the wolf. Pimm, S.L. & Jenkins, C., 2005. Sustaining the Variety of Life. *Sci. American* **293. No.3**, 66-73. The Natural Resources Defense Council is leading campaigns to restrict sonar, to save whales and dolphins, in < www.nrdc.org>.

31. Walters, M.J., 2003. *Six Modern Plagues and how we are causing them* (Washington, D.C.: Island Press).

32. For Sarabhai's early days in the Indian space program, see N. Mathews & F. Morring, Jr, 2004, Dare to Dream. *Aviation Week & Space Technology* **161, No.** 20, 51-52. The background of the Indian space program is described by President Abdul Kalam, 1999. *Wings of Fire.* (Hyderabad: Universities Press).

33. Cohen, J.E., 2005. Human Population Grows Up. *Sci. American* **293, No. 3**, 48-55.

34. See, however, Daly, H.E., 2005. Economics in a Full World. *Sci. American* **293, No. 3**, 100-107.

CHAPTER 14

1. Defending Human Rights Worldwide, *Human Rights Watch*, 24 Oct. 2001; <www.hrw.org>. Cockburn, A., 2003. 21st Century Slaves. *Nat. Geogr.* **204, No. 3**, 2-29. Bales, K. 2002. The Social Psychology of Modern Slavery. *Sci. American* **286, No. 4**, 80-89. Basu, K., 2003. The Economics of Child Labor. *Sci. American* **289, No. 4**, 84-91.

2. See Ref. 18 of Ch. 13.

3. See Note 2 of the Introduction.

4. Pers. comm., Gail Eisnitz, 2001.
5. Eisnitz G. A., 1997. *Slaughterhouse, the shocking story of greed, neglect and inhumane treatment inside the U. S. Meat Industry* (Amherst, N.Y.: Prometheus Books). The World Society for the Protection of Animals has as Web site <www.wspa.org.uk>.
6. *Nucleus: The Magazine of the Union of Concerned Scientists* **23, No. 1**, Spring 2001, <www.ucsusa.org/member/ucsdialog.html>. Additional data are in <www.keepantibioticsworking.org>.
7. *Nucleus* **22, No. 4**, winter 99-00, 6.
8. *U.S. News World Report* **123, No. 8**, 22 – 24, 1997.
9. Lyman, H. F., with Merzer, G., 1998. Mad Cowboy. Plain Truth from the Cattle Rancher Who Won't eat Meat. (New York: Scribner). Regarding the eating of brains, see Rhodes, R., 1998. Deadly Feasts. The "Prion" Controversy and the Public's Health. (New York: Touchstone). Also see Prusiner, S.B., 2004. Detecting Mad Cow Disease. *Sci. American* **291, No. 1**, 86-93.
10. <www.jeffreywigand.com/doj.php>.Public Hearings for the Framework Convention on Tobacco Control, Statement of Jeffrey S. Wigand, 13 Oct. 2000;
11. See Note 2 of the Introduction.
12. <www.annals.org/> is a Web site on screening and treatment for depression. Kapadia, R.I., 1992. *Primer of Universal Healing, a new approach to coronary heart disease.* (J.T. Desai, Navajivan Madranalaya, Ahmedabad, 380 014, India; Girish Modi, 908 Magnolia Dr., Enola, PA, 17025). The Web site of the International Association of Yoga Therapy is in <http://iayt.org/>. Stix, G., 2003. Ultimate Self-Improvement. *Sci. American* **289, No. 3**, 44-45. Gage, F.H., 2003. Brain, Repair Thyself. *Sci. American* **289, No. 3**, 46-53. Holloway, M., 2003. The Mutable Brain. *Sci. American* **289, No. 3**, 78-85. Caplan, A.L., 2003. Is Better Best? *Sci. American* **289, No. 3**, 104-45. Tsien, J.Z., 2007. The Memory Code. *Sci. American* **297, No. 1**, 52-59.
13. Mellon, M. & Fondriest, S., 2001. Data on how much antibiotics used in livestock agriculture are not available to the public. *Nucleus* **23**, No. 1, 1-3; <www.ucsusa.org/act>.
14. Turning Cows into High-Tech Milk Machines. *Humane Farming Association, Special Report*, P.O.Box 3577, San Rafael, CA, 94912.
15. Fish poison is in <www.gotmercury.org>.
15. Torpedo for the Seabed Treaty? *TIME*, July 19, 1982, 38.
17. Powering the Next Millennium, will the U.S. wake up to the new dawn of renewable energy? *Greenpeace Mag.* **4, No. 2**, 1999. Parfit, M., 2005.

Future Power. Where will the World get its next Energy Fix? *Nat. Geographic* **208, No.1**, 2-31.

18. Wind Energy - Clean Energy for our Environment and Economy, *Am. Wind Energy Assoc.* 16 Oct. 2001, <http://www.awea.org>.

19. Renner, M., 2001. U.S. Contempt for Alternatives. *Worldwatch* **14, No. 5**, 2 & 11. *Sci. American* **295, No. 3**, 46-114, 2006 **is** a Special Issue on Energy's Future: Beyond Carbon. How to Power the Economy and Still Fight Global Warming.

20. Ogden, J. M. 2002. Hydrogen: The Fuel of the Future? *Physics Today*, April 2002, 69-75. Ashley, S., 2005. On the road to fuel-cell cars. *Sci. American* **292, No. 3**, 62-69. Wald, M.L., 2007. Is Ethanol for the Long Haul? *Sci. American* **296, No. 1**, 42-49. Gates, B., 2007. A Robot in Every Home. *Sci. American* **296, No. 1**, 58-67.

21. Mattick, J.S., 2004. The Hidden Genetic Program of Complex Organisms. *Sci. American* **291, No. 4**, 60-67. Bains, W., 1998. *Biotechnology, from A to Z.*, 2nd ed. (Oxford Univ. Press).

22. Ast, G., 2005. The Alternative Genome. *Sci. American* **292, No. 4**, 58-65. Genome Sequencing, *Nat. Center for Biotech. Information*, 3 Oct. 2001. I. Dunham, I, with more than 200 co-authors, 1999. DNA sequence of human chromosome 22. *Nature* **402**, 489-495; a meter-long sequence is reproduced in <www.ncbi.nlm.nih.gov/genome/seq/>. Collins, F.S. & Barker, A.D., 2007. Mapping the Cancer Genome. *Sci. American* **296, No. 3**, 50-57.

23. Church, G.M., 2006. Genomes for All. *Sci. American* **294, No. 1**, 46-55. The Ecological Risks and Benefits of Genetically Engineered Plants. *Science*, 15 Dec. 2000; also see *Nucleus* **22, No. 3**, 2000. Sweeney, H.L, 2004. Gene Doping. *Sci. American* **291, No. 1**, 62-69.

24. The July/August 2002 issue of *Worldwatch* is largely devoted to the risks of the rush to Human Genetic Engineering, and the larger agenda of the human biotech industry. Also see Schatten, G., Prather, R. & Wilmut, I., 2003. Cloning Claim is Science Fiction, Not Science. *Science* **299, 5605**, 344. On the other hand, a new institute is dedicated to the cloning of human embryonic stem cells, in Lehrman, S. 2003. Terms of Engagement. *Sci. American* **289, No. 1**, 82-83

25. A nicely illustrated overview of stem cells, discussing also the politics involved, is in Weiss, R., 2005. The Power to Divide. *National Geographic* **208, No. 1**, 2-27, and in The Future of Stem Cells *Sci. American* **293, No. 1, A1.**, 2005. Also see Clarke, M. F. & Becker, M. W., 2006. Stem Cells: The Real Culprits in Cancer? *Sci. American* **295, No. 1**, 52-59.

26. Heath, J.R. & Ratner, M.A., 2003. Molecular Electronics. *Phys. Today*

56, No. **5,** 43-49. Hutcheson, G.D., 2004. The First Nanochips. *Sci. American* **290, No. 4,** 75-83. Seaman, N.C., 2004. Nanotechnology and the Double Helix. *Sci. American* **290, No. 6,** 64-75. Roco, M. C., 2006. Nanotechnology's Future. *Sc. American* **295, No. 2,** 39. Schafmeister, C.E., 2007. Molecular Lego. *Sci. American* **296, No. 2,** 76-82B discusses design and manufacture of nm-scale structures. Bruno, G., 2007. Carbon Nanonets spark new Electronics. *Sc. American* **296, No. 5,** 77-83, provides an overview of organizations working on carbon nanonet materials.

27. Rees, M., 2003. *Our Final Hour. A Scientist's Warning how Terror, Error, and Environmental Disaster Threaten Humanity's Future in This Century – On Earth and Beyond.* (New York: Basic Books).

28. The conditions in the concentration camp are described by survivors Béon, Y. 1997. *Planet Dora. A Memoir of the Holocaust and the Birth of the Space Age.* (Boulder, CA.: Westview Press), and Zembsch-Schreve, G. 1996. *Pierre Lalande, Special Agent.* (Barnsley, U.K.: Pen & Sword Books Ltd.). The psychology of inmates is studied by Frankl, V. E., 1963. *Man's Search for Meaning.* (New York: Pocket Books). I was asked by the editors of *Nature* to write the von Braun history in Gehrels, T., 1994. Of truth and consequences. *Nature* **372,** 511-512. Since that time I have received additional testimony from survivors of Dora, how von Braun witnessed and perhaps even caused at least one of the hangings of a group of inmates. We must take into account that it was dangerous to collect evidence, because any prisoner who looked at him would be killed. The German view of the Dora history is in two volumes by Stuhlinger, E. & Ordway, F.L., 1994. *Wernher von Braun, Crusader for Space.* (Malabar, Florida: Krieger Publ. Co.). von Braun's brother Magnus was also at Dora.

29. The Sputnik-1 event and what it played is heard on <www.hq.nasa.gov/office/pao/History/sputnik/>. Squabbles to keep women from space flight are in Nolen, S., 2003. *Promised the Moon: the Untold Story of the first Women in the Space Race* (New York: Four Walls Eight Windows).

30. *Aviation Week & Space Technology,* 25 Jan. 2001. The Journal has an overview of spacecraft in a January issue each year, and it is remarkable complete in keeping its readers informed on activities at NASA and on all aspects of spaceflight.

31. Lewis, J. S., Matthews, M.S. & Guerrieri, M.L. eds. 1993, *Resources of near- Earth.* (Tucson: Univ. of Arizona Press). Whipple, F. *et al.,* 1973. The 1973 report and recommendations of the NASA Science Advisory Committee on Comets and Asteroids. NASA **TMX-71917.** The mission to Eros is in <http://near.jhuapl.edu/>.

<www.jpl.nasa.gov/solar_system/asteroids_comets/asteroids_missions.html> gives a general overview of comet and asteroid missions.
32. The Journal of The Planetary Society is *The Planetary Report.*.
33. Oberg, J., 2003. China's Great Leap Forward. *Sci. American* **289, No. 4**, 76-83.
34. Poison Mars. *Aviation Week & Space Techn.* **165, No. 6**, 21, 2006.
35. Parker, E.N. 2006. Shielding Space Travellers. *Sc. American* **294, No. 3**, 40-47.
36. <http://nmp.jpl.nasa.gov/ds1/tech/ionpropfaq.html> has data on ion propulsion. Also see Ref. 4 for Ch. 4 regarding ion propulsion for the DAWN spacefraft. A report on the Japanese mission is in Mecham, M., 2005. Quick Kiss *Aviation Week & Space Techn.* **163, No. 18,** 42.
37. Scott, W.B., 2004. To the Stars. *Aviation Week & Space Techn.* **160, No.9**, 50-53, also available as <www.AviationNow.com/awst>.

CHAPTER 15

1. Reviews of astrobiology are in Jones, B. W., 2004. *Life in the Solar System and Beyond.* (New York: Springer Verlag); and Lunine, J., 2005. *Astrobiology: A Multidisciplinary Approach.* (San Francisco: Addison Wesley). The geometries in the text are demonstrated in <http://library.thinkquest.org/C003763/index.php?page=planet08>. See also <http://exoplanets.org/>,<http://listes.obs.ujf-grenoble.fr/wws/info/exoplanets>. Appenzeller, T., 2004. Search for Other Earths *Nat. Geographic* **206, No. 6,** 68-95. Proposals for spacecraft systems are described in Morning, F., Jr., 2002. Mirror Technology Push aids hunt for new Earths. *Aviation Week & Space Technology*, **157. No. 6,** 44-48.
2. Lissauer, J. J. , 1999. How common are habitable planets? *Nature* **402**, C11-14.
3. Details of Frank Drake's pioneering, SETI and UFOs are in the Journal *Mercury*, 21, 114–143, 1992; <www.planetarysystems.org/drake_equation.html> has an updated version. Also see Tipler, F.J. Extraterrestrial intelligent beings do not exist. *Phys. Today,* April 1981, 9 & 70-71, which was debated in *Phys. Today,* March 1982, 26-38. The conclusion, that there may be life but that there is an unspecified low chance of advanced life, is also reached by Ward, P. D. & Brownlee, D. 2000. *Rare Earth. Why Complex Life is uncommon in the Universe* (New York: Springer Verlag).
4. See Note 4 of Ch. 13.
5. The words "opposition effect" were used first for an asteroid, in Gehrels, T., 1956. Photometric studies of asteroids. V. The lightcurve and phase

function of 20 Massalia *Astrophys. J.*, **123**, 331-338.
6. The collision on the Tenerife runway is described in the reference of Note 2 for the Introduction; 583 people died in a runway collision for which the causes were complex but in part due to having visitors in the cockpit of an American plane.
7. <www.seti.com>

CHAPTER 16.
1. The Earth Charter. Values and Principles for a Sustainable Future, with a brochure available from <www.earthcharter.org>. A Special Issue of *Sci. American*, "Crossroads for Planet Earth" begins with Musser, G., 2005. The Climax of Humanity. *Sci. American* **293, No. 3**, 44-47, containing an Action Plan for the 21st Century. The Special Issue ends with Gibbs, W.W., 2005. How Should We Set Priorities? *Sci. American* **293, No. 3**, 108-115.
2. Viewpoint of K. Gottfried, chairman of the Board of the Union of Concerned Scientists, *Nucleus* **22**, No. 4, winter 1999-00 issue (no page number).
3. Swanson, G.J., 2004. *America the Broke* (New York: Double Day).
4. Barnett, T.P.M., 2004. *The Pentagon's New Map. War and Peace in the Twenty-first Century.* (New York: G.P.Putnam's Sons). Much more extreme is <www.aynrand.org>.
5. Sakharov, A., 1990. *Memoirs* (New York: Alfred A. Knopf).
6. Reid, T.R., 2004. *The United States of Europe* (New York: Penguin).
7. <www.interpol.int>
8. <www.deoxy.org/sciwarn.htm> The 2001 quote was prepared for the 100th anniversary of the Prizes. Nobel Laureate Nicolaas Bloembergen (pers. comm., 2002) estimated that the concept statement was mailed to about 250 Laureates, of whom 104 signed it.
9. Renner, M. 1997. *Fighting for Survival: Environmental Decline, Social Conflict, and the New Age of Insecurity.* (New York: W.W.Norton&Co.). Ayres, E., 1999. *God's Last Offer. Negotiating for a Sustainable Future.* (New York: Four Walls Eight Windows). Renner, M. 2002. Lessons of Afghanistan. Understanding the Conditions that Give Rise to Extremism. *World·Watch* **15, No. 2**, 36-38.
10. Foot, M.R.D., 2001, *S. O. E. in the Low Countries* London, U.K.: St. Ermons Press). 11. Note 2 of the Introduction. Two colleagues of Postma were also to be killed, but they were overlooked and left in their cell.
12. See Note 4 of Ch. 13.
13. <www.globalexchange.org/economy/rulemakers/topTenReasons.html>, and <www.wto.org/>.

14. Mann, P., 2001. "Bin Laden – One of Many Suspects." *Aviation Week & Space Technol.* **155, No. 12**, 38, Sept. 17. Also see Chapman, C.R. & Harris, A.W., 2002. A Skeptical Look at September 11th. How We Can Defeat Terrorism by Reacting to It More Rationally. *Skeptical Inquirer* **26, No. 5**, 29-34.; <www.csicop.org/si/2002-09/9-11.html>.

15. Ehrlich, A. & Ehrlich, P., 1987. *Earth* (New York: Franklin Watts), p. 248. Lovins, A.B., 2005. More Profit with Less Carbon. *Sci. American* **293, No. 3**, 74-83. Strom, R. 2007. *HOT HOUSE. Global Climate Change and the Human Condition.* (New York: Springer Verlag).

16. < www.reliefweb.int/ >

17. < www.cartercenter.org >

18. <www.savethechildren.net/homepage/>

19. Bullard, C.W., 1988. Avoid the Pitfalls that line the Road to Effective Activism. *Audubon Activist*, January-February 1988, 11. Regarding fraud, I do not accept any requests by phone, and in any mailing I check that a Board of known proponents has endorsed it. This is often not the case. The mailing may look impressive and professional, registration with authorities is shown, the scheme is legal and it is possible that something useful is being done. But that it is a shyster's sham can be recognized by the fact that the letter comes from only one person who signs impressively as President, or Executive Officer or the like. You can be sure that this person draws an income from the gullible donors; these methods are taught in books on how to become a millionaire.

20. www.iicdmi.org/about.htm, also see www.nationalhomeless.org, and <www.propoor.org>

21. <www.assetindiafoundation.org> <www.akanksha.org> Sachs, J.D., 2005. Can Extreme Poverty be eliminated? *Sc. American* **293, No.3**, 56-65. Bardhan, P., 2006. Does Globalization Help or Hurt the World's Poor? *Sci. American* **294, No. 4**, 84-91.

22. Kalam, A. P. J. A., with Tiwari, A., 1999. *Wings of Fire* (Hyderabad: Universities Press).

23. We got as far as supporting one Center for Akanksha by members of the Lunar and Planetary Laboratory of the University of Arizona and family with the support of the University of Arizona Foundation.

24. Jamison, K.R., 2004. *Exuberance: the Passion for Life* (New York: Alfred A. Knopf). A classic of joy is in Yutang, L., 1938. *The Importance of Living* (New York: Reynal and Hitchcock); it also describes a combination of Eastern and Western thought.

ACKNOWLEDGMENTS OF PEOPLE

The entire text has been critiqued at various stages by Eric Chaisson, Aleida Gehrels, Jo-Ann Gehrels, Derrick Gentry, Sig Kutter, Therese Lane, Shirley Marinus, Travis Metcalfe, Brian May, Jennifer O'Brien, Kevin Prahar, Steve Shawl, and Shawn-rika Van Ess. I received help with specific topics from Jean Barrett, John Cocke, Lyn Doose, George Gehrels, Neil Gehrels, Linda Gregonis, Shermali Gunawardena, Carl Koppeschaar, Leonid Ksanfomality, David Kring, Brian Marsden, Lisa Martin, Joe Montani, Maria Schuchardt, Jim Scotti, Jeff Seligman, Fred Spier, Barbara Timmermann, Wieslaw Wisniewski, Milo Wolff, and an anonymous referee. Sanya Glisic made all the drawings, with the expert advice of Ellen McMahon; preparations had been made by Jennifer Evans and Jodi Johnston. Special acknowledgment is due to ZENIT, the magazine of the highest quality for information, in the Netherlands. I thank you all for your support, which was essential for my education and for making the book.

ACKNOWLEDGMENTS FOR ILLUSTRATIONS

Fig. 2.3, http://Photojournal.jpl.nasa.gov, PIA00827, Dr. R. Albrecht, ESA/ESO Space Telescope European Coordinating Facility; NASA
Fig. 2.4, NASA, AS12-48-7135
Fig. 2.5, http://Photojournal.jpl.nasa.gov, PIA03104, NASA/JPL-Caltech
Fig. 2.6, http://Photojournal.jpl.nasa.gov>, PIA00108
Fig. 2.7, NASA/JPL
Fig. 4.2, http://Photojournal.jpl.nasa.gov, PIA00136
Fig. 4.3, http://Photojournal.jpl.nasa.gov, PIA02873, NASA/JPL/University of Arizona
Fig. 4.4, http://Photojournal.jpl.nasa.gov, PIA6193, NASA/JPL/Space Science Institute
Fig. 4.5, http://Photojournal.jpl.nasa.gov, PIA00032
Fig. 4.6, http://Photojournal.jpl.nasa.gov, PIA01492
Fig. 4.8, http://Photojournal.jpl.nasa.gov, PIA08438, NASA/ JPL-Caltech
Fig. 5.1, http://photojournal.jpl.nasa.gov, PIA03149;
Fig. 5.5, CCD image made by Robert Gendler, M.D., an eight-frame mosaic using the 12.5" RC, STL11000
Fig. 5.6, http://hubblesite.org/newscenter/archive/releases/1999/26, The Hubble Heritage Team (AURA/ STScI/ NASA)
Fig. 6.3, http://Photojournal.jpl.nasa.gov/, PIA04213, NASA, Space Telescope Science Institute
Fig. 6.4, http://Photojournal.jpl.nasa.gov, PIA08787, NASA/ JPL-Caltech
Fig. 6.5, as for Fig. 5.6
Fig. 6.6, http://Photojournal.jpl.nasa.gov, PIA07911, NASA/ JPL-Caltech/SSC
Fig. 6.7, © Roger Smith/NOAO/AURA/NSF,
Fig. 14.2 was drawn by Maurice de la Pintière, Dora inmate number 3115, it is reproduced with his permission.

GLOSSARY
(Consult also the Index.)

The glossary terms are adapted to this book for college level. Sources are Hopkins, J., 1980. *Glossary of Astronomy and Astrophysics* (Univ. Chicago Press); Purves, W.K., Sadava, D., Orians, G.H. & Heller, H.C., 2004. *Life: The Science of Biology,* 7th ed. (Salt Lake City: W.H. Freeman Co.); *Merriam-Webster's Collegiate Dictionary,* 2002, 10th ed., (Springfield, Mass: Merriam-Webster, Inc.); and Chaisson, E.J., 2006. *Epic of Evolution. Seven Ages of the Cosmos* (New York: Columbia Univ. Press).

Angstrom (Å)	angstrom = 10^{-8} cm
AU	(see Astronomical Unit)
aeon	10^9 years
AIDS (Acquired Immune Deficiency Syndrome)	a virus destroying white blood cells so that the victim becomes subject to opportunistic diseases
alga (*plural* **algae**)	one-cell or multi-cellular water plants
Allende	a well-known meteorite, a large carbonaceous chondrite which fell near Allende, Mexico in 1969
amino acids	constituents of proteins containing the (NH,) amino (COOH) and carboxyl groups of compounds. Building blocks of proteins
amplitude	the extent of a vibratory movement measured from the mean position to an extreme
Ångstrom	Anders Jonas (1814-1874).
angstrom	a unit of length of 10^{-8} cm
anthropic	(Fr. anthropos = human being) relating to human beings
anthropic principle	the idea that the universe is the way it is because we are here to observe it; the universe is made for us
aphelion	point on a planetary orbit farthest from the sun

asteroid	also called **minor planet**. A moving object of stellar appearance, without any trace of cometary activity (but note that the same definition may also apply to a satellite or to an exhausted cometary core).
Asteroid Belt	a region of space, lying between the orbits of Mars and Jupiter
astrobiology	the study of the origin, evolution, and distribution of past and present life in the universe; also known as bio-astronomy or exobiology
astronaut	someone trained to take part in space flight
Astronomical Unit (AU)	for which the length is the mean distance between Sun and Earth, 1.496×10^8 (remember AU as 1.5×10^8 kilometers)
astrophysics	the science that deals with the physical properties of astronomical objects
atom	unit of ordinary matter that is thought to be made up of a tiny nucleus surrounded by electron shells
ATP	adenosine triphosphate, an organic molecule that acts as energy currency in life forms
aurora	light radiated by ions in the earth's atmosphere, mainly near the geomagnetic poles, stimulated by bombardment by energetic particles ejected from the sun (see solar wind)
aurora borealis	for the northern lights; **aurora australis** occurs in the southern hemisphere
Australopithecus	early hominid of 4-3 million year ago; approximately of the same size and build as a chimpanzee, but with shorter arms and ability to walk on two legs
bacteria	single-celled, fungus-like microorganisms
baryons	primarily protons and neutrons
Big Bang	"Big Bang" theory has the universe starting from all matter at one small confinement, followed by an expansion

GLOSSARY

binary star	a pair of stars in orbit around each other
biosphere	the part of the earth, atmosphere and mantle in which forms of life exist
black hole	a superdense body with gravitation so strong that even light cannot escape
blue-green algae	simplest algae; a group of simple organisms like bacteria
Bohr, Niels	1885-1962, Danish physicist
Brahe, Tycho	1546-1601, Danish astronomer
brane	a string particle, named after membrane
brown dwarf	a star of the lowest luminosity and low mass, bordering those of the largest planets
c	speed of light in a vacuum, 2.99792458×10^5 km/sec (remember it as 300,000 km/sec)
carbohydrate	a sugar, or a sugar like organic substance
CCD	an electronic charge-coupled device for imaging
cell	one of the complex units making up living things, capable alone or in interaction with other cells, of performing fundamental functions of life, and forming the smallest structural unit of living matter that functions independently
cell theory	all organisms consist of cells, and all cells come from preexisting cells
centimeter (cm)	0.01 meter; for conversions, remember that 1 inch =2.54cm
cerebellum	part of the brain concerned with muscle coordination
cerebral cortex	the layer of gray matter that covers the cerebrum and performs the highest mental processes
cerebrum	largest part of the human brain, that performs the higher functions
chemical bond	electrical attraction holding atoms together within a molecule

chlorophyll	the green coloring matter of plants that facilitates synthesis from light; it is made up of molecules like $C_{55}H_{72}MgN_4O_5$ and $C_{55}H_{70}MgN_4O_6$
chondrite	a stony meteorite usually characterized by the presence of chondrules
chondrule	small spherical grain varying from microscopic size to the size of a pea, usually composed of iron, aluminum, or magnesium silicates. They may have formed at the time of solar-system formation.
chromosome	threadlike packet of DNA and proteins found in the nucleus of every cell. They contain the hereditary factors
clade	all of the organisms descended from a particular common ancestor
clay	a soil constituent of small particles
clone	identical organisms produced from a common ancestor by asexual means
closed universe	a model universe that stops expanding at some time in the future, after which it contracts to a point much like that from which it began
COBE	Cosmic Background Explorer, a spacecraft in NASA's Explorer series
coma	nearly spherical region of diffuse gas, some 150,000 km in diameter, surrounding the nucleus of a comet
comet	a diffuse body of solid matter and gas, which orbits the sun. The orbit is usually highly elliptical and inclined to the ecliptic. Some comets are seen crashing into the sun
comets, family of	an aggregation of comets with similar aphelion distance near Jupiter's orbit; "Jupiter's family of comets."
complementarity	a term used for example for an electron, which can be a particle, but that it can also have a wave form

consciousness	being aware of something within oneself
convection	transfer of heat by motion of gas or fluid between regions of unequal density, which are due to non-uniform heating
Copernicus, Mikolaj	1473-1543, Polish astronomer
corona	like a crown, the solar corona may look like a low-density gas extending for millions of kilometers, with temperatures of 10^6 K
cosmic abundance	about three-quarters hydrogen, one-quarter helium with traces of the other elements
cosmic evolution	a grand synthesis of all the changes in the assembly and composition of radiation, matter, and life, throughout history
cosmic rays	charged particles, with energies ranging between 10^6 and 10^{20} eV, but mostly near 10^9 eV. They are mostly protons, some come from the sun, but the origin of most cosmic rays is still unknown
cosmology	the study of the large-scale structure, evolution and origin of the universe
cosmonaut	someone trained in the Soviet Union (now primarily Russia) to take part in space flight
creationism	the position that the account of creation in the first book of the Bible, "Genesis", is literally true
dark energy	name given to the unknown cause of acceleration in the expansion of intergalactic space
dark matter	unseen mass in galaxies and galaxy clusters whose existence is only been inferred indirectly and has not been observed
Darwin, Charles Robert	1809-1882, English naturalist
decoupling	the event at age 400,00 when atoms were completed, after which photons moved freely in space, allowing

	matter and radiation to behave differently
degenerate matter	matter that is degenerate in the sense of its pressure being a function of density only, independent of the temperature
density	a measure of compactness
differentiation	the process of the separation by gravity of materials that have different weight
DNA	deoxyribonucleic acid, the inheritance material consisting of long molecular chains present in the chromosomes of all living matter
Doppler shift	displacement of spectral lines in the radiation received from a source due to its relative motion along the line of sight. Sources approaching the observer are shifted toward the blue (shorter wavelengths); those receding are shifted toward the red. For cosmology, the expansion of space (not objects in space) is measured, and relativistic effects are taken into account
elementary particle	a basic component of atoms
elliptical galaxy	a galaxy having an elliptical or spherical shape, some more than others, and composed mostly of old stars and little interstellar matter
eccentricity, e	of an elliptical orbit. It is the amount by which an orbit deviates from being circular, in terms of 0 being a circle and 1 being a straight line
ecliptic	plane of the earth's orbit (with minor irregularities smoothed out). It is seen as a line, in a map of the sky, along which the sun and the major planets appear to move
ecosphere	the regions of the earth capable of supporting living organisms
ecosystem	a community where the relationship between organisms and their environment is regarded as a unit

Einstein, Albert	1879-1955, American (German-born) physicist
electrodynamics	the science that deals with motion of electrical charges under the influence of electric and magnetic fields
electron	a stable, negatively charged particle
ethic	a system of principles to live by
eon	or aeon, 10^9 years
eV	electron Volt, unit of photon energy equal to the energy of an electron accelerated from rest by 1 volt. The wavelength associated with it is 12.398 Angstrom and the velocity of an electron is about 580 km/sec
event horizon	a region within which no event can ever be seen or known by anyone outside, also termed the "surface" of a black hole
evolution	a process in which our whole world is a progression of interrelated phenomena
fallout	usually radioactive material resulting from a nuclear explosion coming down through the atmosphere
fermi	a unit of length of 10^{-13} cm
Fermi, Enrico	1901-1954, American (Italian-born) physicist
fission	a nuclear reaction that releases energy when heavy nuclei break down into lighter ones
flagella	threadlike appendages of plants or animals
fossil	anything found in the layers of Earth that is recognizable as remains or traces of an organism from a former geological age
frequency	the number of times a repetitive event occurs in a given unit of time.
fungus	any of a major group of lower plants that lack chlorophyll; they include rusts, mildews, smuts, mushrooms, and usually bacteria

fundamentalism	a movement or point of view marked by rigid adherence to basic principles
Fundamentalism	a Protestant movement characterized by a belief in the literal truth of the Bible
fusion	the process in which two nuclei collide and coalesce to form a single, heavier nucleus
g	Gram
G	the gravitational constant; Newton's gravitational constant
Galilean satellites	the four largest satellites of Jupiter: Io, Europa, Ganymede, and Callisto
Galilei, Galileo	1564-1642, Italian astronomer and physicist
gamma rays	photons of the highest frequency group (wavelengths shorter than a few tenths of an Angstrom, energies greater than 100 eV); there is no discrete cutoff between gamma-rays and X-rays. Usually gamma-rays come from the nucleus and X-rays come from the inner electron shells
gene	a segment of any DNA molecule containing information for the construction of one protein, hence responsible for directing inheritance from generation to generation
gene mapping	determining the relative positions of genes on a chromosome
gene therapy	replacing the diseased gene with a healthy copy
genetics	a branch of biology that deals with the heredity and variation of organisms
genome	the DNA sequence of the genes of an organism
globalization	the process or policy of making something worldwide in scope or application
glucose	simple carbon form of sugar that occurs widely in organisms

gravity wave	the gravitational analog of an electromagnetic wave whereby gravitation radiation is emitted at the speed of light from any mass that undergoes rapid acceleration
Gyr	giga-year = 10^9 yr
h	the Planck constant
habitable zone	a three-dimensional region of "comfortable" temperatures that surrounds every star
Halley's comet	its orbit was computed by Edmund Halley in 1704, at which time he predicted that the bright comet of 1682 would return in 1758. Appearances occurred in 1910 and 1986
Hawking, Stephen	British astrophysicist
Heisenberg, Werner	1901-1976, German physicist
holism	our world is seen in terms of interacting wholes that are more than the sum of their components
Homo erectus	species of hominid appearing about 1.6 Myr ago, which stood completely upright, erect, but their skulls, jaws and teeth were not fully modern
Homo Sapiens	modern human beings, "wise man."
hot spot	a place where volcanism occurs on the earth's surface due to a plume of hot material that rises through the mantle. In contrast to most other volcanism, hot spots do not necessarily occur along plate boundaries. Hawaii and Yellowstone are good examples of hot spots that occur in oceanic and continental plates, respectively, far from plate boundaries
Hoyle, Fred	1915-2001, British astrophysicist
HR diagram	Hertzsprung-Russell diagram: in present usage, a plot of brightness against temperature for a group of stars. Related plots are for the color and the spectrum. The main spectral types are O, B, A. F, G, K, M

hydrocarbon	a compound containing only hydrogen and carbon
i	inclination of the orbit, which is the angle between the plane of the orbit and the plane of the ecliptic
IAU	International Astronomical Union
IGM	intergalactic medium
igneous rock	solidified rock from a molten or partially molten state
inch	= 2.54 cm.
indeterminancy	a combined expression for wave/particle duality and the uncertainty principle (see below)
Industrial Revolution	the rapid change in world economy, begun in England in the late 18th Century, marked by the introduction of power-driven machinery
inorganic compound	a substance that contains any combination of elements except carbon and hydrogen
interstellar grain	a conglomerate of molecules present in between the stars
invertebrate	an organism without a backbone
ion	an atom or group of atoms that carries a positive or negative electric charge as a result of having gained or lost one or more electrons
ionization	the removal of an electron from an atom
IR	infrared light of wavelengths longer that those of visible radiation
ISM	interstellar medium
isotope	atomic nucleus with the same number of protons but different numbers of neutrons (e.g. deuterium is an isotope of hydrogen)
IUM	inter-universal medium
K	degrees Kelvin, a scale of temperature. The unit is the same as of the centigrade degree, but **absolute zero** is at zero Kelvin, 0 K, which is –273.16 degrees centigrade

Kepler's laws	1. Each planetary orbit is an ellipse with the sun at one focus. 2. When a planet is closer to the sun, it is attracted stronger, so it travels faster in that orbit. A **law of areas**: summarizes that equal areas are swept out in equal times by a line between planet and Sun. 3. a **harmonic law** summarizes that the square of the period about the sun is proportional to the third power of the average distance of the planet
kinetic energy	energy associated with motion. $E = \frac{1}{2} m v^2$ where m is the mass and v the (non-relativistic) velocity
km	kilometer = 10^5 cm.
Leonardo da Vinci	1452-1519, Florentine genius: painter, sculptor, architect and engineer
lichen	growth made up of an algae and fungus on a solid surface such as a rock
lightyear	the **distance** that light travels in one year in a vacuum, 9.46×10^{12} km (remember it as 10^{13} km; it is 6.3×10^4 AU)
luminosity	total radiant output of a star, per second
Mach, Ernst	1838-1916, Austrian physicist and philosopher
Mach's principle	a pre-relativity statement to the effect that the local frame of reference is determined by some average of the motion of all the matter in the universe. In essence, Mach's principle says that space, which is the arena in which matter interacts, is itself an aspect of that matter
magma	molten material underneath the earth's crust from which igneous rock is formed by cooling
magnitude	logarithmic system of measuring brightness, initiated by the Greeks. An astronomical magnitude is proportional to $-2.5 \log_{10} I$, where I is the intensity. Each magnitude step is about a factor of 2.5 **fainter.** The sun is near magnitude -26, the brightest stars are near zero, the faintest visible ones near +6, and the

	faintest observed with the largest telescopes near +26
main sequence	the principal sequence of stars in the HR diagram, containing more than 90% of the stars we observe, that runs diagonally from the upper left (high temperature, high luminosity) to the lower right (low temperature, low luminosity)
Malthus, Thomas Robert	1766-1834, English economist
mantle	the interior of a planet, between the crust and the core
mass	a quantity of matter
meiosis	the division of a cell nucleus, whereby the chromosomes do not divide, but are equally shared by the two new (sex) cells
Mendel, Gregor Johann	1822-1884, Austrian botanist
metabolism	the sum of all chemical reactions that energetically support a living organism, starting from energy sources that are either chemical (environmental nutrients) or physical (sunlight)
meteor	a "shooting star," the streak of light in the sky produced by the transit of a meteoroid through the earth's atmosphere; also the glowing meteoroid itself. The term "fireball" is sometimes used for a meteor approaching the brightness of Venus; the term "bolide" for one approaching the brightness of the full moon
meteorite	extraterrestrial material which survives passage through the Earth's atmosphere and reaches the Earth's surface as a recoverable object or objects
micron	10^{-4} cm = 10^{-9} km
mile	1.609 km
mineral	in most cases it is a crystal, which are the building blocks of rocks

mitochondria	small structures in eukaryotic cells that generate their oxidation energy from nutrients
mitosis	the division of a cell nucleus into two equal parts, in which all the chromosomes divide equally, yielding two cells identical to the first
molecule	the smallest particle of a substance that retains the properties of the substance and is composed of atoms
multiverse	all universes
mutation	a permanent change in the structure of DNA
Myr	10^6 yr.
nanotechnology	the use of molecular machines to fabricate or replicate molecular structures
natural selection	the waveforms of a new species and its environment interact and thereby affect each other
neo-Darwinism	a combination of traditional Darwinian evolution and Mendelian genetics, also termed "the modern synthesis."
neuron	a biological, or nerve, cell in a brain
neutrino	an extremely light, possibly mass-less, elementary particle that is affected only by gravity and the weak atomic force
neutron	electrically neutral elementary particle found in the nucleus of an atom, except in hydrogen
neutron star	a star composed primarily of neutrons in a small dimension
Newton's laws	1. A body remains in a state of rest or uniform motion when left to itself. 2. The net force on a body is equal to the product of its mass and its acceleration. 3. When two bodies interact, the force on the first due to the second is equal and opposite to the force on the second due to the first.

nucleus	central region of an atom, containing protons and neutrons, which in turn consist of other particles
nova	a star that suddenly becomes brighter greater and then may fade away again in the following months or years
Ockham's razor	*Entia non sunt multiplicanda* ("entities are not to be multiplied"). A doctrine formulated by William of Ockham in the fourteenth century. Any hypothesis should be shorn of all unnecessary assumptions; if two hypotheses fit equally well, the one that makes the fewest assumptions should be chosen.
Oort cloud	a region extending to about 100,000 AU from the sun; gravitationally it is the storage place for comets that, at that distance, is barely bound to the sun
open universe	a universe that expands forever
organic compound	compound containing carbon and hydrogen
outgassing	losing gases, usually small amounts from unusual places as rocks
ozone	O_3
parsec (pc)	1 parsec = the distance where 1 AU would be seen as one arcsecond = 206,265 AU = 3.26 lightyears
perihelion	the point in the orbit closest to the sun
photon	a quantum of light; the smallest separable amount of energy in a beam of light (see Note 3 of Ch. 8)
photosynthesis	the process in chlorophyll by which plants manufacture food by means of light
Planck, Max Karl	1858-1947, German physicist
Planck's constant	h, is the constant that relates the frequency of a photon to its quantum of energy
Planck mass, length, time	quantities related to the fundamental units of gravity, G; velocity of light, c; and the Planck constant, h. The expressions that follow are not needed for understanding the discussions in this book:

GLOSSARY

Planck mass = $(hc/G)^{1/2}$ = 5.46 x 10^{-5} grams.
Planck length = $(Gh/c^3)^{1/2}$ = 4.05 x 10^{-33} cm
Planck time = $(Gh/c^5)^{1/2}$ = 1.35 x 10^{-43} sec.

Planck's quantum principle Light, or any other electromagnetic wave, can be emitted or absorbed only in discrete quanta, whose energy is proportional to their frequency, E=hf.

planetesimal bodies of intermediate (about 100 m) size, most of which accreted into larger bodies

plankton minute floating organisms found in water, serving as food for many animals

plasma a completely ionized gas in which the temperature is too high for atoms as such to exist and which therefore consists of free electrons and free atomic nuclei

protein one of a group of nitrogen-containing organic compounds of large molecular size, of one or more chains of amino acids; protein is an important part of living material, performing a wide variety of activities in the cell

proton a positively charged elementary particle; the nucleus of a hydrogen atom

pulsar an object which has the mass of a star and a radius no larger than that of Earth and which emits radio pulses with a very high degree of regularity (periods slow down with age; they range from 0.03 sec for the youngest to more than 3 sec for the oldest). Pulsars are understood to be rotating, magnetic neutron stars, the end products of supernovae

quantum the indivisible unit in which waves may be emitted or absorbed (see photon)

quantum physics physics based on quantum theory which imposes certain fundamental limitations on measurements of physical quantities and leads in many cases to a discrete behavior (in contrast to classical, or pre-quantum physics, which predicted a continuous behavior).

quark basic building block of protons, neutrons, and other elementary particles

quasar an object with a dominant star-like component, and its emission line spectrum showing a large redshift. The light of most, if not all quasars is variable over time intervals between a few days and several years, so their diameters must not be much larger than the diameter of the solar system; yet they are the intrinsically brightest objects known. The basic problem of quasars is that they emit too much radiation in too short a time from too small an area.

radioactivity the spontaneous disintegration of unstable atomic nuclei

Reciprocity Principle stating that an optical system, and many situations in nature, are reversible and reciprocal

redshift the lengthening or stretching of lightwaves coming from a source moving away from us

refractory capable of withstanding high temperatures

regolith the layer of fragmental rocky debris produced by meteoritic impact on the surfaces of satellites, asteroids or the moon

relativity, special, general theories originated by Albert Einstein. The special theory (1905) is based on the postulates that if two systems are in relative motion with steady velocity, it is impossible for observers in either system to learn more than that it is in relative motion. Also that measurements of the velocity of light in either system, regardless of the position of the source of light, always give the same numerical value. It leads to equivalence of mass and energy and to the increase of mass with increased velocity. The general theory (1915) is an extension of the special theory, explaining the force of gravity and related phenomena in terms of the curvature of four-dimensional space-time.

resonance	when there is a regularly occurring beat, such as when an asteroid executes exactly 3 orbits in the time of 1 Jupiter orbit. Gravitational effects are then periodically enhanced so that the asteroid orbit is changed
RNA	ribonucleic acid, similar to a single strand of DNA with Uracil in place of Thymine
solar wind	a radial outflow of energetic charged particles from the solar corona
species	a class of objects or individuals grouped by virtue of their common attributes and assigned a common name (see Sec. 7.1)
spectral types, B, A, etc.	see "HR diagram"
spin	a quantum number describing an internal property of elementary particles that is related but not identical to the everyday concept of spin
star identification	bright variable stars have letters, for example AZ Orionis (AZ Ori). Most other stars have catalog numbers, as CoD -35 10525 is number 10525 in the zone of −35 degrees declination of the Cordoba Durchmusterung
stem cell	a cell capable of extensive proliferation as in the formation of red blood cells
string	the fundamental one-dimensional object of string theory
strong nuclear force	the strongest of the four fundamental forces (strong and weak, electromagnetic, gravity), holding elementary particles together within protons and neutrons and them together to form atoms
supernova	a gigantic stellar explosion in which the star's brightness increases some 100 million times
symbiosis	the living together, in a mutually beneficial union that aids survival or evolution, of two organisms of

	different species
synergy	a term used particularly in the world of business: the action of two or more components to achieve more of an effect than each is individually incapable
taikonaut	someone trained in the People's Republic of China to take part in space flight
temperature	a measure of the average kinetic energy of the particles of a system (for instance the molecules in a gas)
terrorism	systematic use of violence, intimidation, and overwhelming fear to achieve an end, but not in open warfare
thermodynamics	the physics of the interrelationships between heat and other forms of energy
thermonuclear reactions	reactions that take place when different atomic nuclei are brought together at high temperatures, resulting in changes of structure (breakup or fusion) of participating nuclei accompanied by a release or absorption of heat
Trojans	Trojan asteroids occur in two groups preceding and following Jupiter in its orbit. The groups have on average about the same distances from Sun and Jupiter
T Tauri star	eruptive variable sub-giant star associated with interstellar matter and believed to be in the process of gravitational contraction
Uncertainty Principle	The uncertainty in measuring the position of a particle varies inversely with that of its momentum (product of mass and velocity). It follows that an atomic or nuclear process can not be measured without affecting the process.
universal evolution	the general term for the ensemble of all observed evolutions, from the earliest phases of our universe's beginning until the present time

Universal Time (UT)	the local mean time of the meridian of Greenwich, England. UT is essentially Greenwich Mean Time, counted from 0 hr at Greenwich mean midnight
Van Allen belts	two doughnut-shaped belts in the earth's magnetosphere (at about 1.4-1.5 and 4.5-6.0 Earth radii) where energetic charged particles from the solar wind are trapped by Earth's magnetic field
UV	ultraviolet
virus	submicroscopic particle that can reproduce only within a host cell
Wallace, Alfred Russel	1823-1913, English naturalist
wavelength	of a wave is the distance between two adjacent troughs or crests
wave/particle duality	the concept in quantum mechanics that there is no distinction of waves and particles, and that each may behave like the other
waves	variations that transfer energy from point to point in a medium such as space
weak nuclear force	a weak fundamental force, operating over a very short range notably in radio-activity (see strong force)
white dwarf	a star of high surface temperature, low luminosity, and high density (10^5-10^8 g cm^{-3}), with roughly the mass of the sun but radius of the earth, that has exhausted most or all of its nuclear fuel, believed to be a star near its final stage of evolution
world view	a comprehensive understanding of the universe and of humanity's relation to it
X-rays	radiation of wavelengths between about 0.1 and 100 Angstroms; more energetic than ultraviolet light, but less than gamma-rays

INDEX

A

Absolute Zero, 62, 119, 141, 228, 296
ACBAR. *See* Arcminute Cosmology Bolometer Array Receiver
Aceh, 194
Achieving Sustainable Social Equality through Technology (ASSET), 256
acid rain, 199ff
acidity, 40, 199
Acquired Immune Deficiency Syndrome (AIDS), 204, 287
acquired variations, 5
Active Galactic Nuclei (AGN, 97, 102
adaptive optics, 94
addiction, to smoking, 211
adenosine triphosphate (ATP), 288
advertising, role of, 195, 202, 214
Africa, human evolution and, 4, 166, 201, 205, 262
Agent Orange, 190
agriculture, 199, 217, 278
    pesticide use, 198ff, 211, 219
    small farms, 209ff, 217, 219, 245
Ahmedabad, i, xiv, 198, 206, 213, 252ff, 259, 278
AIDS. *See* Acquired Immune Deficiency Syndrome
aircraft, 15, 20, 82, 217
    altitudes, 16, 18ff, 38ff, 202
    opposition effect seen from, 234, 281
    supersonic, 202
Akanksha, 5, 256ff, 283
Al Qaeda, 187
Alabama, 181, 274
alcohol, 217
algae, 150, 161ff, 287, 289, 297
Allende meteorite, 24ff, 143, 287
altruism, 247
amino acid, 157ff, 231, 287, 301
Amnesty International, 208

amoebas, 161
amphibians, 163
Amsterdam, 208
Andaman Islands, earthquake, 15
Andromeda Galaxy/Nebula (M31), 87, 91, 93, 97, 174, 175
angstrom, 287
Ångstrom, Anders Jonas, 287
animal rights, 208
animals
    and food industry, 214
    evolution of, 17, 68, 109, 152, 154, 161ff, 183, 203, 206, 209ff, 214, 218, 220, 246, 273, 293, 301
Antarctica, 24, 94, 120, 201
anthropic principle, 88, 99, 100, 102, 110, 127, 287
Anti-ballistic Missile Treaty, 249
antibiotics, in meat industry, 161, 209, 214, 278
anti-particles, 118
aphelion, 47, 287, 290
Apollo missions, 30ff, 226ff
Apophis, 179
Arcminute Cosmology Bolometer Array Receiver (ACBAR), 94, 120
Arizona, i, xii, 19, 21, 37, 39, 83, 94, 95, 129, 178, 180, 197ff, 200, 231, 280ff
    copper industry, 194, 200
    Glen Canyon Dam, 197
Arizona Radio Observatory, 83, 231
arms race, 187, 233
arms sales, 195
Armstrong, Neil, 225ff
arthropods, 163
ASSET. *See* Achieving Sustainable Social Equality through Technology
asteroid belts, vi, 45ff, 59, 146, 177ff
Asteroid Tugboat, 274
asteroiders, 266

asteroids, viiff, xii, 23, 26ff, 35, 38, 45ff, 58, 64, 69, 87, 145ff, 154, 175ff, 181, 203, 216, 224ff, 243, 266ff, 273ff, 280ff, 302, 304
    impacts, 10, 30, 33, 36, 58, 141, 147, 149, 150, 164, 175ff, 233, 273
astrobiology, 231ff, 281, 288
astronauts, 31, 87, 203, 225, 288
astronomical unit (AU), 26, 47, 59ff, 143, 287ff, 297, 300
Astronomy Picture of the Day, 261
atmosphere(s), vii, 8, 10, 15ff, 30ff, 50ff, 65ff, 83, 92ff, 113, 131, 141, 149ff, 160, 175, 180ff, 197ff, 224, 226, 230, 234, 238, 265ff, 288ff, 293, 298
atmospheric pressure, 16, 39
Atom Bomb Museum (Hiroshima), i, 178, 185, 188, 258
atomic bombs. *See also* nuclear weapons, 178, 186
atoms, i, viii, xiii, 17, 50, 66ff, 78, 88, 92, 97, 108ff, 137ff, 167, 182ff, 220ff, 238, 262, 288ff, 296, 299, 301, 303
    properties of, 131
AU. *See* astronomical unit
Aurora Australis, 10, 288
Aurora Borealis, 10, 288
Australopithecus, 166, 288
autobiography of Darwin, 107, 110ff, 151ff, 205, 213, 230, 272, 291
averted vision, and turning head degrees, 23
Awash River, 165

B

bacteria, 110, 150, 157, 160, 209, 288ff, 293
Bains, 279
balloon, 19, 120, 228
Bangkok, 207ff, 256
Banks, 151
Barlow, 272
barred spiral galaxy, 86ff, 97
baryon, 98, 129
Beagle, HMS, 151, 272ff

Bean, 31
Beethoven, xi, 257
belief systems, 4
Béon, i, 280
BepiColombo spacecraft, 33
Beppo-SAX spacecraft, 95
Betelgeuse, 71
BGH. *See* Bovine Growth Hormone
Big Bang, 114ff, 182, 268ff, 272ff, 288
Big Dipper, Big Bear (constellation), 71
Big History, 108, 269, 273
bin Laden, 187, 252ff, 283
binary code, 238, 240ff
binary stars, 73
bio-diversity, 203
biogenetic material, on other planets, 231
biological warfare, 189ff, 245
biomass, energy from, 217
biosphere, 149, 197, 272, 289
biotechnology, 168, 218
birds, 154
birth control, 6, 198, 205, 252
black holes, 35, 78ff, 89, 91ff, 113, 174, 268, 289, 293
Black-Hole Finder Probe, 95
Bloembergen, 282
blowback, 194ff
blue-green algae, 161, 289
blueshift, 174, 230, 268
Bohr, 123, 270, 289
BOOMERANG balloon telescope, 120
Borrelly, Comet, 227ff
Bovine Growth Hormone (BGH), 214
Brahe, 129, 265, 271, 289
brain, evolution of, 39, 130, 164ff, 195, 246ff, 289, 299
brane, 289
Brazil, 203
bromines, 202
brown dwarf, 230, 289
Brownlee, 265, 281
Bruno, 55, 95, 280
businesses, globalization of, 167, 186, 191, 216, 232ff, 246ff, 255, 273, 304

C

California, Gulf of, 11, 14, 95ff, 178, 269
Callisto, 55ff, 294
Cambrian Explosion, 155ff
Canada, 200, 203
cancer, 188, 201, 208, 211, 215, 219
Canis Minor, 76
carbon, 18, 39, 73, 111, 121ff, 136ff, 149ff, 156ff, 161, 187, 200, 280, 294, 296, 300
carbon dioxide, 18, 39, 149, 156ff, 200
carbon monoxide, 137, 150
Carson, 264
Carter, 276
Carter Center, 256
Cascades Mountains, 15
Cassini, 56
Cassini spacecraft, 56
caterpillar effect, 136
cattle production, 203, 209ff
Caucasians, 4
CCD, 285, 289
CDI. *See* Center for Defense Information
cells, xiii, 108ff, 122, 131, 150, 155ff, 160, 162, 183, 189, 216ff, 279, 287, 289, 298ff
Celsius, A., 16, 154, 263
Centaurs, vii, 28, 58ff, 146
Center for Defense Information (CDI), 187, 189, 275
centipedes, 163
Central Intelligence Agency, 194
centrifugal effect, 7, 133, 135, 144
cerebellum, 165, 289
cerebral cortex, 116, 165, 205, 211, 213, 215, 247, 289
cerebrum, 213, 289
Ceres (minor planet), 46ff
Cerro Tololo International Observatory, 91
Chaisson, i, 107, 261, 269, 284, 287
Chandra spacecraft, 95, 108, 268
Chandrasekhar, 94, 99, 108, 117
charged particles, magnetic fields and, 9ff, 23, 176, 291, 303, 305
Charles, Prince of Wales, 222, 274
Charon, 29, 59, 267
chemical warfare, 189
Chernobyl, 188
chicken manure, 195, 210
chicken production, 195, 210, 214
Chicxulub crater, 180ff
children, 5, 16, 43, 154, 183, 192, 200, 206ff, 220, 238, 254ff,
Chile, 91, 94, 195
chimpanzees, 166
China, 30, 187, 193ff, 202, 211, 213, 226, 248, 281, 304
Chiron, 48, 59, 267
chlorine, 189, 202
chlorophyll, 161ff, 290, 293, 300
chondrite, chondrule, 287, 290
chordates (Phylum Chordata), 163
Christian, 96, 108, 269
chromosomes, 108ff, 157ff, 183, 217ff, 244ff, 292, 298ff
Churchill, 246, 250ff
clade, 290
clams, 162
clear cutting, 12, 203
Clementine spacecraft, 37
clenbuterol, 208
climate change, 40
cloning, clones, 220ff, 279, 290
clouds, 17, 21, 33, 40, 51, 60ff, 81ff, 94, 111, 119ff, 133ff, 174ff, 200, 229, 231, 234
clusters of galaxies, 93
cobalt, 216, 224
Cold War, 48, 178, 186, 193ff, 226, 246, 251
collective action, 256
collisions, 13, 24, 38, 46, 49, 51, 55ff, 66, 87ff, 113, 135, 137, 143, 145ff, 156, 174ff, 225, 232, 282
colonialism, 194
Colorado River, 11ff, 197ff, 264
coma, 46, 62ff, 146, 290
Coma Cluster, 93

Comet Schwassman-Wachmann, 63
Comet Wild, 82
comets. *See also* trans-Neptunian
    objects, viff, 23, 26ff, 35, 38, 41, 46ff,
    58ff, 82, 98, 119, 144ff, 174ff, 224,
    227ff, 243, 267, 273ff, 280ff, 290, 300
communication with extraterrestrial life,
    127, 235, 238, 298
complementarity, 290
complexity, 83, 107ff, 155, 157, 164,
    183, 201
Comprehensive Test Ban Treaty, 249
computer modeling, 136
computers, 61, 87, 135ff, 168
concentration camp, i, 189, 222ff, 280
condensation nuclei, 141
cones, of the eye, 17, 74
conglomerates, 191ff, 255, 296
consciousness, 126, 291
conservation, 197ff, 216, 258
Constellation X Observatory, 95
constellations, 23, 76, 91, 93, 145
Constitution of the World, 243
consumers, 209ff, 214, 219
continental crust, 12, 14
continents, plate tectonics and, 3, 12, 36,
    39, 42, 87, 231ff, 264
convection, 7ff, 67, 291
Convention on Climate Change, 40
Cook, 151
Copenhagen, 270
Copernicus, 291
copper industry, 194, 200
corals, 162, 204
corona, 68, 291, 303
cosmic abundance, 118, 291
Cosmic Background Explorer (COBE),
    119ff, 290
cosmic dust, 265
cosmic evolution, 291
cosmic rays, 92, 291
cosmic symphony, i, 114, 132
cosmic winds, 264
cosmology, 95, 97, 269, 291ff
cosmonauts, 226
cosmos, 94, 112, 183

Coyne, i
Crab Nebula, 77
crabs, 163
craters, impact, 30ff, 58
creationism, 291
Cretaceous-Tertiary extinction, 180
critical mass, 101, 117
Cronkite, 276
cruelty, 207ff, 256
C-type asteroids, 48
Cuban missile crisis, 186
Curie, 123
curvature of space, 21, 26, 116, 302
cyclic universe, 113
CYCLOPS, 236

D

da Vinci, 297
dark energy, 88, 99, 120, 269ff, 291
dark matter, 84, 88, 98, 120, 291
Darwin, 107, 110ff, 151ff, 205, 213,
    230, 272, 291
    on Galápagos Islands, 151
    The Origin of Species, 153, 272
    The Voyage of the Beagle, 272
DASI. *See* Degree Angular Scale
    Interferometer.
DAWN, 49, 266, 281
death, viii, 77, 109, 161, 173, 175, 178,
    182ff, 204, 209, 218, 222
decoupling, 291
Deep Impact mission, 63, 267
Deep Space I craft, 228, 231
deep-sea vents, 14
degenerate matter, 176, 292
Degree Angular Scale Interferometer
    (DASI), 120
Deimos, 35
Delsemme, 272
Denmark, wind energy in, 216
density, 7, 12, 28, 37ff, 50, 67, 78, 81,
    102, 111ff, 133, 135ff, 143, 147,
    176ff, 291ff, 305
density waves, 136

deoxyribonucleic acid (DNA), xiii, 108, 156ff, 165, 168, 184, 217ff, 227, 273, 279, 290ff
　　recombination, 10, 102, 159, 217
Department of Defense (U.S.), 196
depression, 153, 213, 278
desalinization, 205
detection, 46, 231, 274
deterrence, 186, 275
Diana, Princess, 198
Dickens, 167
differentiation, 147, 181, 224, 292
dimensions, in string theory, 114, 173, 303
dinosaurs, 46, 164, 181
diplomatic corps, 196
Dirac, 123
dirty snowballs. *See* comets, 61, 146
diversification, 255
diversity, global, 196ff, 256
DNA sequence, 279, 294
dolphins, 203, 277
Doppler, 96, 292
Doppler effect, Doppler shift, 292
Dora, 222ff, 280, 285
doubt, 114, 180ff, 202, 225, 237
Drake, 231ff, 281
droplet haze, on Venus, 40
drugs, use and abuse of, 208, 214, 218
dust storms, on Mars, 40, 42, 154, 227
dust tail, 63
dust, space. *See also* interstellar grains, vi, 21, 23, 30, 32, 39ff, 61ff, 73ff, 81ff, 98, 135ff, 144, 147ff, 174ff, 180, 227
Dutch East Indies, 250
dwarf planet, 30, 46

E

Earth, viff, 3ff, 14ff, 20, 25ff, 32ff, 45ff, 68, 71, 79, 82, 94, 109, 112, 119, 129, 138, 147ff, 154, 159ff, 174ff, 184, 186, 203ff, 222ff, 230ff, 241, 243ff, 254, 264ff, 272ff, 280ff, 288, 293, 298, 301, 305

Earth Charter, 243, 282
earthquakes, 14ff, 264
Eartth
　　surface of, 56
East African Rift Valley, 3, 165
eccentricity, 47, 292
eclipses, 27, 68, 230
　　solar, 28, 68, 92
ecliptic plane, 24, 28, 45, 60, 134, 144, 146, 229ff
ecology, 197
economics
　　globalization and, 167, 191ff, 244, 248, 251, 256, 294
　　of poverty, 5, 207, 243, 246, 251ff
ecosphere, 197, 254, 292
ecosystem, 215, 292
Ehrlich, 254, 283
Einstein, 92ff, 100, 123, 129ff, 293, 302
Eisenhower, ix, 190ff, 246, 249, 255, 258, 275
　　on military-industrial complex, 191ff, 255
El Niño, 201
electrodynamics, 293
electromagnetic effects, 8, 50, 66
electromagnetic quantum vacuum, 117, 228
electromagnetism, 67, 129
electron waves, 124, 130, 238
electronic communication, 5, 168, 193
electrons, 8ff, 23, 38, 50, 56, 63, 66ff, 78, 93, 113ff, 136ff, 167, 176, 216, 228, 238, 270, 288, 290, 293ff, 301
　　behavior of, 56, 114, 121ff, 140, 238, 288ff
elements, formation of, 58, 69, 72, 76, 95, 99, 101, 125, 138ff, 174, 291, 296
elliptical galaxies, 135, 174
elliptical orbits, 26, 37, 47, 57ff, 130, 292
empire(s), 194ff, 244ff, 275
Encke, Comet, 64
Encyclopedia, 158
endangered species, 203, 276ff
Endeavour, HMS, 151

energy, 14, 17, 51, 65ff, 78, 88, 91ff, 99ff, 115ff, 130, 137ff, 149, 154, 167, 175ff, 183, 185ff, 197ff, 203ff, 215ff, 221ff, 269ff, 279, 288, 291, 293, 297ff
   generation of, 69
   in quantum theory, 114ff, 126, 228, 301
   wind, 9, 23, 33, 40ff, 55ff, 67, 141, 216, 288, 303, 305
environment, xii, 4, 16ff, 23, 93, 109ff, 130ff, 139, 142, 151ff, 161ff, 183ff, 191, 197ff, 210, 215, 221, 228, 232, 248, 255ff, 265, 276, 279, 292, 299
environmental awareness, 197
eons, 150, 293
Eros (asteroid), 227, 281
erosion, 11, 36, 201
escape velocity, 79, 222
ethanol, 217
ethics, 191, 251
   of global responsibility, 191, 251
eukaryotic cells, eukaryots, 299
Eurasian Plate, 13, 15
Europa, 55ff, 267, 294
Europe, 11, 74, 111, 141, 154, 190, 193, 210, 234, 244ff, 253, 263, 265, 282
   united, 11, 74, 111, 141, 154, 190, 193, 210, 234, 244ff, 263, 265, 282
European Space Agency (ESA), 226, 231, 285
European Union, 204, 209
evaporation, 39, 143, 148, 161, 176, 201, 205, 228
event horizon, 293
evolution, i, viff, 3, 12, 14, 17, 25, 36, 39, 46, 50, 59, 79, 83ff, 89, 101, 107ff, 129, 132, 135ff, 150ff, 166ff, 173, 181ff, 197, 205, 214, 218, 229, 231ff, 236ff, 243, 246ff, 261ff, 272ff, 287ff, 299, 303ff
   interstellar grains, 23, 81ff, 108, 112, 141ff, 155, 182
   of solar system, 146
   of universe, xiii, 101, 173ff, 182, 232, 243
exchange effect, 5

exercises, 182, 210ff
   stress-reduction, 211, 213, 237ff, 241, 254
exponential notation, 6
extinction, 164, 180ff, 203, 247
extraterrestrial life
   communicating with, 127, 235, 238, 298
   predictions of, 127, 235, 238, 298
   search for, 127, 235, 238, 298
exuberance, 283
eyes, 11, 17, 21, 23, 31, 35, 70, 84, 91, 122, 132, 159, 163, 201, 211, 213
   exercises for, 213

F

Fahrenheit, G., 16, 263
fall-out, 187ff, 293
family planning, 206
FDA. *See* Food and Drug Administration
federation, 248
Fermi, 293
Festou, 267
Feynman, 125, 128, 270ff
fighting for survival, xi, 46, 101, 108, 117, 153ff, 165, 167, 204, 245ff, 303
finches, 109, 151ff
fire, control of, 39, 166ff, 202, 252
fish, 161, 163, 188, 214, 272
fission, 185ff, 215, 293
FitzRoy, 151
flagella, 293
Flanders, chemical warfare in, 189
flatworms, 162
Flessa, 264
food, 3, 5, 109, 122, 152, 154, 161ff, 182, 184, 198, 204ff, 228, 300ff
Food and Drug Administration (FDA), 209, 219
food industry, 214
fossils, 163, 200, 216, 293
Framework Convention on Tobacco Control, 278
France, 187, 189, 222
Frankl, 280

freedom fighters, 251
French Revolution, 228
Freon, 202
frequency, 19, 126, 130, 175, 177, 293ff, 300ff
frost, 37, 40
    water frost on Moon, 37
fuel cells, 217
fundamentalism, 191, 245, 294
fungi, 156, 161, 293, 297
fusion generators, 215
fusion, nuclear, 137ff, 187
future, i, xiff, 5, 14, 30, 56, 87, 107, 151, 173, 177, 183, 187, 191ff, 200ff, 215, 222ff, 236, 243, 245, 247, 252ff, 290

G

Gagarin, 225
Gaia hypothesis, 150, 272
galactic plane, 83, 87ff, 134ff, 144
Galápagos Islands, 109, 151ff
galaxies
    formation of, 120, 174
    spiral, 89, 135, 136
galaxies (*See also* Milky Way galaxy), viiff, xii, 71, 79, 81, 84ff, 109, 113ff, 135ff, 173ff, 182ff, 230, 232, 268ff, 291
Galilean satellites, 55ff, 294
Galileo, 48, 53ff, 294
Galileo spacecraft, 55
gamma radiation, 19, 175
gamma rays, 19, 89, 93, 139, 185, 294
gamma-ray bursters (GRB), 92, 95
gamma-ray bursts, 79, 92, 95, 268, 273
Gandhi, 182, 206
Ganymede, 55ff, 294
gas lungs, 189
gases, compression of -. *See also* atmosphere(s), 20, 26, 37ff, 50, 61ff, 122, 149ff, 181ff, 189, 204, 217, 277, 300
Gell-Mann, 114, 270
gene doping, 219
gene mapping, 294

gene therapy, 294
genes, 158, 183, 218ff, 294
genetic engineering, 199, 217, 219
genetic variation, 112
genetics, 157, 168, 294, 299
genome sequencing, 279
genomes, 217, 279, 294
geodynamo, 264
germ warfare, 189
Germany, 222
germline engineering, 219
germs, 275
geysers, on Triton, 57
glaciers, 201, 204
Glen Canyon Dam, 197
Glenn, 276
Gliese, 175
global responsibility, i, xii, 246
global warming, 20, 163, 181, 200ff, 243ff, 257ff, 276
globalization, 167, 191ff, 244, 248, 251, 256, 283, 294
globular clusters, 73, 87ff
glucose, 294
glycine, 231
Gold, 272
Goldilocks Principle, 43
Goodall, 276
Gorbachev, i, xii, 276
Gottfried, 282
government contracts, 191
Graham, Mount, telescopes on, 94
grains, 23ff, 58ff, 81ff, 89, 123, 130, 136, 139, 140ff, 214
    interplanetary, 23ff
    interstellar, 23, 81, 108, 112, 141ff, 155, 182
Grand Canyon, formation of, 12, 35
graphite, 141
gravitational lensing, 92
gravity, 6, 12, 29, 49ff, 72ff, 88, 92ff, 115, 117, 129, 133, 137, 144ff, 177, 222, 227, 229, 268, 292ff
Gravity Probe B, The Relativity Mission, 95
GRB. *See* Gamma ray bursters

Greene, 114, 131, 270
greenhouse effect, 20, 42, 69, 150, 200
Greenland, magnetic North Pole near, 9
Greenpeace, 200, 279
Gross, 271
Ground Zero, 185
gunpowder, 167, 255
Guth, 269

H

habitable zone, 232, 281, 295
half-life, 188
Halley, Comet, 62, 64, 295
Hawaii, 15, 35, 94, 178, 295
    telescopes in, 15, 35, 94, 178, 295
Hawaiian islands, 15
Hawking, 268, 276, 295
Hayabusa explorer, 228
hazards, 175ff, 188, 213, 254, 274
health, 10, 200, 204, 209, 214, 221, 256
    lifestyle and, 214ff
Hecuba (asteroid), 47
Heisenberg, 123, 125, 131, 270, 295
Hektor (asteroid), 47
helium, 49, 50, 65, 77, 101, 110ff, 133ff, 150, 174, 186ff, 291
    formation of, 186
Herschel space telescope, 94
Hertzsprung-Russell (HR) diagram, 74, 295
HFA. *See* Humane Farming Association
Himalaya Mountains, 14, 36, 94, 201
    Indian National Observatory in, 94
Hiroshima, atomic bomb at, i, 178, 185, 188, 258
history, big, i, xiii, 12, 25, 40, 55, 57, 87, 107ff, 123, 145, 147, 153ff, 182, 192ff, 222, 243, 245, 275, 280, 291
Hitler, 182, 192, 246, 250
holism, 111, 214, 295
hominids, evolution of, 165
Homo erectus, 295
Homo sapiens. *See* humans
Horsehead Nebula, 82
hot spot, 15, 36, 295

Hoyle, 99, 101, 269, 295
HR Diagram. *See* Hertzsprung-Russell (HR) diagram
HST. *See* Hubble Space Telescope
Hubble, 90ff, 227, 285
Hubble Space Telescope (HST), 94ff, 227
Human Genome Project, ix, 217
human rights, 196
Humane Farming Association (HFA), 208, 278
humans (Homo sapiens), i, ix, xii, 9, 14, 20, 39, 56, 110, 115, 122, 125, 131, 156ff, 177ff, 195, 204, 209ff, 236, 239, 246ff, 277ff, 295
    genetic engineering of, 199, 217, 219
    overpopulation by, 204ff, 233
hurricanes, 52, 201, 276
Hussein, 253
Huygens, 32, 53, 56
Huygens Probe, 53, 56
hybrid cars, 216
hydrocarbon, 56, 296
hydrogen, 49, 65, 69, 73, 77, 82, 95, 101, 110ff, 150, 157, 174, 186, 217, 238, 247, 291, 296ff
    as fuel, 217
hydrogen bombs, 186, 247
hygroscopic, 141

I

IAU. *See* International Astronomical Union
Icarus (asteroid), 47, 272
ice fields, 201
Ida, (asteroid), 48
IGM. *See* intergalactic medium
igneous rock, 296
IMAGE spacecraft, 264
immigrants, cultural impact of, 244
impact features, 36
inclination, 27, 47, 59, 64, 296
independence movements, 192dd
indeterminacy principle . *See also* uncertainty principle, 110, 125

India, i, xii, 30, 94, 187, 191ff, 226, 248, 252ff, 278
    Akanksha program in, 5, 256, 283
Indian National Observatory, 94
Indian Ocean, earthquake under, 15
Indian Plate, 13
Indonesia, 194, 203, 251ff
Industrial Empire, 194
Industrial Revolution, 167, 202, 296
industry, industrial complex
    globalization of, 20, 167, 187ff, 245, 250, 256, 275, 279
industry, industrial complex. *See also* military-industrial complex, 191ff, 255
inflation, 99, 116, 119
Inflation Probe, 117
inflation theories, 116, 119
infrared (IR), 18ff, 41, 83ff, 145, 162, 296
Inquisition, Galileo Galilei and, 55
insects, 152, 163
Institute for Food and Development Policy, 198
INTEGRAL mission, 95, 120
interaction, 10, 35, 63, 82, 92ff, 111ff, 131, 139, 149, 156, 230, 248, 289
intergalactic medium (IGM), 93, 143, 296
intergalactic space, expansion of, 86ff, 102, 173, 291
INTEGRAL. *See* International Gamma-Ray Astrophysics Laboratory
International Association of Yoga Therapy, 278
International Astronomical Union (IAU), 29, 145, 175, 265, 296
International Gamma-Ray Astrophysics Laboratory (INTEGRAL), 95, 120
international relations, 195, 249
International Save the Children Alliance, 256
International Seabed Treaty, 216, 278
International Space Station (ISS), 226
interplanetary grains, 23
Interpol, 196, 249, 251ff

interstellar clouds, 89, 111, 136ff, 175, 229
interstellar gas, 82, 87, 136, 139
interstellar grains, dust, 23, 81ff, 108, 112, 135, 141, 155, 182
interstellar medium (ISM), 82, 93, 109, 139, 141ff, 151, 231, 296
interstellar molecules, 83, 123, 141
inter-universal medium (IUM), 117, 142, 296
invertebrates, 162, 296
Io, 51ff, 294
ion propulsion, 49, 228, 235, 266, 281
ion tail, 63
ionization, 156, 296
ions, 20, 27, 49, 63, 227ff, 266, 281, 296
IR. *See* infrared
Iraq, 189, 195
Irian Jaya, 194
iridium, 180, 224
irregular galaxies, 90, 135
Islam, in Indonesia, 253
ISM. *See* interstellar medium
isolation, 3, 152, 244
isolationism, 193
isotope, 296
Israel, 187, 195
    nuclear weapons in, 187, 195
ISS. *See* International Space Station
Itokawa (asteroid), 228
IUM. *See* inter-universal medium

J

Japan, 15, 185, 193, 195, 210, 228, 274
jellyfish, 162
Johnson, C., 193, 195, 196, 198
Johnson, L.B., 225
Joint Dark-Energy Mission, 95
Jupiter, vii, 27, 45ff, 64ff, 130, 145, 177, 227ff, 266, 274, 288, 303
    and Comet Shoemaker-Levy 9, 27, 274
    magnetosphere, 58, 305
    satellites of, 29, 146, 227, 294

K

Kalam, 257, 277, 283
Kamchatka, 15
Kant, 143
Kapadia, 213, 278
Kapitza, 261
Kauffman, 269
Kelvin, 7, 228, 263, 296
Kelvin, Lord (William Thompson), 263
Kennedy, 211, 225
Kepler, 26, 129, 230, 265, 271, 297
Kepler's Laws, 129
Kieffer, 265
Kilauea Crater, 15
Kitt Peak National Observatory (Arizona), 83, 178
Kring, 274, 284
Ksanfomality, 266, 284
Kuiper Belt, 59, 64, 267
Kutter, 158, 273, 284
Kyoto Protocol, 202

L

La Niña, 201
Laika, 224
LAGEOS. *See* Laser Geodnamics Satellite
Lamb, 271
landmines, 198
Large Binocular Telescope, 94
Laser Geodynamics Satellite (LAGEOS), 237
Laser Interferometer Space Antenna, 95
Laser Interferometric Gravitational Wave Observatory, 95
lava basins, on moon, 31, 147
laws, xiv, 101, 118, 129, 141, 196, 210, 261, 297
leeches, 162
lichens, 297
life, i, viiiff, 3, 10, 20, 25, 28, 35ff, 61, 77, 102ff, 122, 127, 128, 138, 145ff, 173ff, 197, 204ff, 227ff,
   and death, 243
   evolution of, 3, 46, 109, 150, 232, 248
   life expectancy, 4, 184
light, xi, 4, 11, 17ff, 20ff, 27, 33, 38, 40, 52, 54, 63, 67, 69, 70ff, 78ff, 90, 92ff, 116, 119, 124, 130ff, 161, 176, 183ff, 216, 229ff, 258, 271, 288ff,
   speed of, 71, 79, 85, 96, 116, 136, 235, 289, 294
lightyears, 71, 143, 174, 297
livestock production, 209ff, 214, 278
lobbyists, 191
lobsters, 163
Local Group of Galaxies, 93
Loihi Seamount, 15
Lovelock, 150, 272
Lowell, 29, 39, 95
Lubarsky, 208
luminosity, 71ff, 140, 289, 297ff
lunar eclipses, 28
Lunine, 262, 272, 281

M

M 31 (Andromeda Nebula), 87, 91, 97, 174
Mach, 123, 131, 271, 297
MAD. *See* mutually assured destruction
Mad Cow disease, 210, 278
Madagascar, 203
Magalhães, 90
Magellanic Clouds, 90
magnetic fields, 8ff, 50, 67, 78, 136, 144, 293
magnitude, 14, 178, 297
main asteroid belt, 47, 59
Main Sequence, 75
Malthus, 152, 205, 215, 221, 298
mammals, 163ff, 81
manganese, 216
Manuvarya, 259
maria, mare, 31, 147
Mariner spacecraft, 32ff
Mars, vii, 6, 10, 26, 33ff, 59, 70, 144ff, 176, 224ff, 265, 281, 288
   surface of, 35, 39

Mars Climate Orbiter, 6
Marshall Plan, 193
mass, 6, 15, 26ff, 50, 65, 72, 78, 88, 92, 98, 100ff, 126ff, 174, 177, 179, 181, 183, 196, 201, 273, 289ff,
massage, 211
Maxwell Montes, 34
McNamara, 251, 274
meat industry, 209, 214
meditation, 213
meiosis, 298
Mendel, 298
Mercury, vii, 26ff, 45, 55, 70, 144, 263, 265, 281
    outgassing on, 30, 37, 46ff, 150, 300
Merkel, 204
MESSENGER spacecraft, 33
Messier catalogue, 91
metabolism, 298
Meteor Crater (Arizona), 180
meteor showers, 24
meteorites, 24, 143, 155, 180, 287, 290, 298
meteoroids, 24, 30
meteors, 24, 180
    fireballs (*See also* meteorites, 24, 79, 298
methane, 54ff, 83, 141, 149, 156, 160, 200
Midway Island, 15
migration, human, 14, 146
military-industrial complex, 190, 191, 192
Milky Way, vii, 24, 81ff, 111, 133, 174
Milky Way Galaxy, vii, 83
Mine Ban Treaty, 198
minor planets. *See* asteroids
MIR space station, 226, 228
missiles. *See* rockets
missing mass, 88, 98
mitochondria, 299
mitosis, 299
Mogadishu (Somalia),, 194
molecular clouds, 60ff, 135, 140, 175

molecules, xiii, 7ff, 37ff, 50, 63, 65, 82ff, 108ff, 131, 136ff, 149, 155, 158, 160ff, 175, 182, 189, 213, 220, 228, 231, 238, 288ff
    interstellar, 83, 123, 141
mollusks, 162
moment scale, 14
Montreal Protocol, 202
Moon, vii, 30, 36, 41, 56, 68, 144ff, 265ff, 280
    eclipses and, 28
    Papuas and, 4
Mori Daiko, 258
mountain building, 14
multiple scattering, 67
multiverse, vi, xi, xii, xiii, 99ff, 127ff, 159, 174, 182, 232, 299
munitions, 188, 257
    radioactive, 147, 185ff, 215, 224, 246, 293
mustard gas, 189
mutation, 299
mutually assured destruction (MAD), 186

N

Nagasaki, atom bomb at, 178, 185, 188
naked eye observations, 23, 71, 74, 82, 90, 141
nanometers, 220
nanorobots, 221
nanotechnology, 168, 220ff, 262, 299
NASA, 59, 95, 100, 179, 225, 231, 254, 268, 274, 280, 285, 290
National Geographic Atlas, 93, 197, 264, 268, 279
National Missile Defense (NMD), 187
National Radio Astronomical Observatory (NRAO), 83
Natural Resources Defense Council, 277
natural selection, 72, 108, 111, 139, 141, 151, 153, 161, 167, 269, 299
Navy, U.S., 151, 203
Neanderthals, 166
near-Earth objects, 177, 227

Nehru, 192, 275
neo-Darwinism, 299
Neptune, vii, 27, 29, 37, 50ff, 70, 145, 266, 272
and Triton, 266
Netherlands, 208, 251, 268, 284
clenbuteral, 208
Netscape, 261
neurons, 299
neutrinos, 299
neutron bomb, 186
neutron stars, 78, 113, 174, 299, 301
New Guinea, 4
New Horizon mission, 38, 59, 268
Nicobar Islands, 15
NMD. *See* National Missile Defense
Nobel Laureates, on terrorism, 249, 252, 282
North America Nebula, 82
North Pole, magnetic, 9
northern lights, 288
nova, 300
NRAO. *See* National Radio Astronomical Observatory
nuclear fusion, 75, 110, 137ff, 144, 156, 176, 187, 215, 294, 304
nuclear weapons, 178, 186, 190, 247
weapons (*See also* biological weapons; chemical weapons, 178, 185ff, 225, 243ff
nuclei, 50, 66ff, 91, 97, 101ff, 130, 135ff, 157, 167, 176, 183ff, 293, 301ff
heavy, 139, 293
numbers, xiii, 5, 18, 25, 28, 47ff, 71ff, 85, 110ff, 139, 146, 160, 176ff, 199, 207, 216, 232, 250ff, 296, 303

O

O and B stars, 75ff, 99, 135ff, 183
obesity, 215
oceanic ridges, 14, 149
oceans, 11ff, 36ff, 53ff, 150, 154, 155, 161, 182, 197, 200, 266
plate tectonics and, 3, 12, 36ff, 87, 231, 264

octopus, 162
Olympus Mons, 35
Oort, vii, 59ff, 89, 119, 146, 174ff, 181, 235, 243, 267, 300
Oort Cloud, vii, 59ff, 89, 146, 174ff, 235
open clusters, 73, 86, 136
open universe, 300
opposition effect, 234, 281
origin(s), 28, 41, 50, 59, 62, 92, 101, 112, 121, 123, 133, 139, 143, 146, 150ff, 161, 195, 234, 272, 288, 291
Orion, 70, 96
Orion Nebula, 71, 96
outgassing, 30, 37, 46, 48, 150, 300
outsourcing, 195, 200, 255
overpopulation, 204ff, 233
oysters, 162
ozone, 17ff, 175, 197, 201, 300
ozone hole, 202

P

Pacific Coast, 15
Pacific Ocean, 15, 201
Pacific Plate, 14
Papuas, 4
parasites, 162
viruses as, 160, 218
particles
and anti-particles, 118
characteristics of, 9, 21, 23, 41, 56, 58, 61, 73, 92ff, 109ff, 167, 174ff, 221, 227, 243, 271, 288ff, 299ff
partisans, 250
peace dividend, 193
peace, dangers to, i, 186, 188, 193, 214, 226, 249, 251ff
penicillin, 161, 208
Pentagon, 195, 275, 282
People's Republic of China. *See* China
periodic orbits, 64
peripheral vision, 23
pesticides, 198, 211, 219
philanthropy, 192
Phobos, 35
photons, 79, 119, 126, 271, 293, 300

photopolarimetry, 271
photosynthesis, 154, 161, 180, 300
Pioneer spacecraft, 27, 53, 58, 237ff
Pittsburgh, 198
Planck, 100, 115, 123ff, 139, 271, 295ff
Planck length (PL), 300
Planck Time (PT), 115ff, 300
Planck's constant, 100, 126, 300
Planck's quantum principle, 126, 301
planetary nebula, 145
planetary rings, 267
Planetary Society, 281
planetesimals, 301
planets. *See also by name*, vii, xii, 8, 24ff, 46ff, 70, 84, 88, 98, 130, 144ff, 177, 182, 224, 229ff, 264, 281, 289, 292
plankton, 161, 301
plate tectonics, 3, 12, 36ff, 87, 231, 264
platinum group, 224
Pleiades, 73
Pleysier, 250
plumes, magma, 57, 149, 231
Pluto, vii, 25ff, 37, 45, 58, 147, 265ff
  and Charon, 29, 59, 267
polar bears, 201
pollution, 17, 56, 131, 163, 187, 198, 200, 203, 217, 227, 233
  air, 199
  global warming and, 20, 163, 181, 200ff, 243ff, 257, 276
polyps, 162
poor man's space program, 25, 250
population, human, 204, 215, 221, 254
Postma, 250, 282
poverty, 5, 207, 243, 246, 251ff
  terrorism and, 186, 250ff, 304
Powell, Lake, 197, 198
pressure, 8, 16, 39, 41, 50, 62ff, 72, 78, 115ff, 137, 145, 150, 175, 203, 215, 248, 292
primates, 165
Procyon, 76, 78, 267
prokaryotic cells, prokaryots, 156
prostitution, 255

protein, 157, 161, 195, 210, 218, 294, 301
protocols, 189, 202, 248
proton, 100, 111, 118, 121ff, 183ff, 301
protostars, 144
PT. *See* Planck Time
pulsars, 77, 240, 301
pulsating universe, 98, 114

Q

quantum black holes, 268
quantum mechanics, 99, 115, 125, 130, 238, 270, 305
quantum physics, 82, 100, 125ff, 243, 301
quantum principle, 126, 301
quarks, 302
quasar, 302

R

radar, 33, 179, 237
radiation, 4, 9, 17ff, 31, 38, 50ff, 62ff, 93, 102, 118ff, 136ff, 150, 174, 178, 185, 187, 201, 216, 218, 227, 236, 238, 271, 291ff
  electromagnetic, 93
  from nuclear weapon, 178, 187
  gamma, 19, 175
radiation belts (*See also* Van Allen Belts), 9, 50, 53
radioactive wastes, 147, 185ff, 215, 224, 246, 293
radioactivity, 302
rainforests, 203, 214
Rainier Mount, 15
Rare Earth, 281
reciprocity, 108, 216, 229, 258
recycling, 182, 216, 258
red giants, 176
redshifts, 100
Rees, 221, 269, 280
refractory, 302
regolith, 30, 32, 302

re-incarnation, 183
rejuvenation, 183
relativity, 91, 100, 115ff, 302
Reliable Replacement Warhead, 187
Remus, 48
Renner, 262, 279, 282
reptiles, 163, 165
resonance, 20, 129, 137, 303
ribonucleic acid (RNA), 303
Richter scale, 14
Rigel, 71
rings, 36, 52, 56ff, 101, 144
riverbed, 39
RNA. *See* ribonucleic acid
robotics, 30, 217, 227, 235
Robust Nuclear Earth Penetrator, 186
rockets, 10, 19, 187, 222ff, 275
    sounding, 10, 222
rods, of the eye, 17, 74
Romulus, 48
Röntgen, 95
Röntgen radiation, 95
ROSAT spacecraft, 95
Rossi, 95
Rossi X-ray timing explorer (RXTE), 95
rotation, 6, 9, 27, 41, 54, 57, 87, 133, 135, 144ff, 232, 238
roundworms, 162
Russia (*See also* Soviet Union), 179, 187, 193, 198, 205, 226, 248, 291
RXTE. *See* Rossi X-ray timing explorer

S

Sagan, 237ff
Sagittarius, 174
Saint Helens, Mount, 15, 150
Sakharov, 48, 246, 282
San Andreas Fault, 14
Sarabhai, M., i, 206, 252, 261, 277
Sarabhai, V., 206, 252
satellites, 28, 35, 38, 48, 52ff, 144, 224, 227, 294, 302
    exceptionally large, 28
    small, 28, 35, 52ff

Saturn, vii, 27, 29, 37, 50ff, 70, 145, 227, 238, 266
    atmosphere, 52
scallops, 162
scarps, on Mercury, 33, 36
scattered disk objects, 59
Schrödinger, 123
sea anemones, 162
sea cucumbers, 162
sea urchins, 162
seabed, 245
search, xiii, 3, 20, 46, 77, 101, 142, 152, 187, 227, 235, 249, 270
Search for Extra Terrestrial Intelligence (SETI), 236, 281
seawater, 200, 205
    desalinization of, 205
sediments, 11
self-gravitation, 143
SETI. *See* Search for Extra Terrestrial Intelligence
sharks, 110, 163
shockwaves
    from supernova explosions, 78, 92, 136ff
Siberia, 180
silicon, 111, 216
Sirius, 76
skates, 163
slavery, 207ff, 256
Slipher, 95, 174
Sloan Survey, 268
smoking, 211
Smolin, 99, 269
Smuts, 111, 269
snails, 162
society, vi, ix, 39, 98, 109, 125, 167, 177ff, 196, 211, 214, 221, 233ff
SOE. *See* Special Operations Executive
solar eclipses, 28, 68, 92
solar energy, 17, 216, 224
solar nebula, 62, 134ff
solar panels, 216, 221, 228
solar storm, 9, 10, 136, 144, 227

solar system, vi, 24ff, 41, 46, 49ff, 82ff, 108, 123, 133ff, 154, 173ff, 216, 224, 227, 237, 240, 266, 302
solar telescopes, 66ff
Solar Terrestrial Relations Observatory (STEREO), 9, 264
solar wind, 9, 23, 56, 67, 141, 288, 303ff
somatic gene modification, 219
sonar, 203, 277
Souken, i
sound waves, 21, 62, 130, 132
South Pole, magnetic, 9, 37
southern lights, 10
Soviet Union
    Cold War, 48, 178, 186, 193, 226, 246, 251
    space exploration, 41
Soviet Union (*See also* Russia), xii, 48, 178, 186, 193, 224, 226, 291
space, xiv, 6, 17, 20ff, 43, 49, 51ff, 69, 77, 79ff, 127ff, 173, 206, 216ff, 254, 257, 264, 268, 277, 280, 288ff
    atomic, 128
    expansion of intergalactic, 86ff, 102, 173, 291
space flight, 223ff, 280, 288, 291, 304
    manned vs. unmanned, 16, 227ff
Space Interferometry Mission, 231
Space Science Series, 261
space stations, 226
space time, 131
space waves, 111, 128, 129, 130
particles (*See also* interplanetary grains, 9, 21, 23, 41, 56, 58, 61, 73, 92ff, 109ff, 167, 174, 176, 183ff, 221, 227, 243, 271, 288ff
spacecraft (*See also by name*), 6, 9, 17, 23, 27, 33ff, 82, 88, 95, 99, 116, 119, 136, 202, 206, 216, 227ff, 264ff, 280ff
Spacewatch, 178
Special Forces, 196
Special Operations Executive (SOE), 250
species, 108ff, 140, 151ff, 174, 180ff, 203, 215, 220, 232ff, 246, 276, 295ff
spectral type, 75ff, 295, 303

spectroscopy, 64
spectrum of solar radiation, 4, 63, 201, 218, 226
speed, 34, 46, 49, 71ff, 85, 96, 116, 136, 147, 174, 177, 224, 235, 289, 294
spiders, 163
Spier, 108, 269, 273, 284
spin, atomic property of, 238, 303
spiral arms, 86, 90, 135
    star formation in, 86, 90, 135
spiral galaxies, 89, 135, 136
Spirit Lake, 150
Spitzer Space Telescope, 63, 94, 268
sponges, 162
spring tides, 11
Sputnik, 224
squid, 162
SSTs. *See* supersonic transports
stability fringe of solar system, 28, 60
Standard Model, 114
Stardust spacecraft, 82
starfish, 162
stars, vii, 8, 17, 24, 28, 46, 51, 60ff, 108, 113, 119ff, 174, 182ff, 215, 224, 229ff, 289, 292ff
    formation of, 85
State of the World, 204, 276
stem cells, 220, 279, 303
STEREO, 9, 264
Strategic Arms Reduction Treaties, 249
stress, 14, 153
string, 114, 131, 173, 289, 303
string theory, 114, 173, 303
subatomic particles . *See also* branes; electrons; neutrons; protons; quarks, 73, 102, 109, 112, 118, 127, 184, 188
structure, xiii, 20, 42, 54, 74, 87ff, 100, 111, 113, 122, 127, 136, 157, 161, 166, 217, 225, 268, 273, 291ff
S-type asteroids, 49
subduction, 14, 36
submarines, 14
sulfur, 41, 56, 111, 149, 200
sulfur dioxide, 41, 200
    acid rain, 199
sulfuric acid, 40ff, 200

sun, 4, 9ff, 41ff, 84ff, 92, 113, 119, 137, 141ff, 161, 173ff, 201, 216, 229ff, 287ff
   death of, viii, 175
Sun Greetings, 212
sunspots, 67
superclusters, 101
supergiant stars, 75
supernovae, 77, 79, 175, 268, 301
supersonic transports (SSTs), 202
surface, 7, 9ff, 48, 51, 56, 65ff, 74, 92, 141ff, 162, 176, 180, 201, 203, 224ff, 238, 293ff
surveying, 178, 244
Surveyor, 31
survival, xi, 46, 101, 108, 117, 153, 154, 165, 167, 204, 245, 249, 303
   evolution and, xi
   of the thoughtful, 246
   through evolution, 117
sweatshops, 195, 207, 255
Sweden, Chernobyl accident and, 188
SWIFT, 95, 268
Swift spacecraft, 95, 268
Sylacauga, Alabama, 274
symbiosis, 303
synergy, 111, 304

T

taikonauts, 226
Tara Umara, tribe in Northern Mexico, 4, 184
Tasmania, 9
tax, 167, 191, 257
tectonic movement. *See* plate tectonics
telescopes, 20, 30, 33, 38, 51ff, 67, 69, 71, 82ff, 116, 120, 137, 178, 222, 236, 297
television, 17, 33, 40, 128, 130, 190, 206, 210, 228, 252ff
Tenerife accident, 282
Terrestrial Planet Finder, 230
terrorism, 186, 250ff, 304
   street, 252
Tethys Sea, 3, 155

Thailand, 13, 207
The Oprah Winfrey Show, 210
thermostat effect, 155
Three-Mile Island (Pennsylvania), 188
tides, 11, 87, 161, 175
   gravitational, 11, 29
time, in relativity theory, 91, 115
Titan, vii, 29, 53ff, 73
Tito, 251
TNOs. *See* trans-Neptunian Objects
tobacco industry, 211
trans-Neptunian objects (TNOs), 59, 145
treaties, 196, 245, 248
Tree of Life, 160
trends, 110, 201
Triton, vii, 55, 57, 146, 266
Trojan asteroids, 47, 304
tsunamis, 179, 194
T-Tauri stage, 145
Tunguska River, 180
Turkana, Lake, 165
Turkish Commandos, 196
Twain, 64

U

UFOs. *See* unidentified flying objects
U.S. Geological Survey (USGS), 264
U.S. Navy, 151, 203
Ukraine, 188
ultraviolet light, 17, 38, 83, 141, 201, 305
uncertainty principle, 125, 296
unidentified flying objects (UFOs), 234
Union of Concerned Scientists, 187, 189, 209, 244, 274, 278, 282
United Nations, xii, 186, 188, 196, 256
United States, 185, 189, 190ff, 226, 244, 248, 258, 276, 282
Universal Evolution, 107
Universal time (UT), 261, 305
universe(s), i, vii, 17ff, 29, 43, 50, 55, 66, 71, 77, 81ff, 112ff, 143, 151, 173ff, 224, 230, 232, 238, 243, 262, 268ff, 281, 287, 290, 297, 299ff
   cyclic, 113

Planck time in, 300
pulsating, 98, 114
University of Arizona, i, xii, 19, 94, 178, 283, 285
unpredictability, 112, 123, 125
Up with Children, 257
Upanishads, 184
upright walk, 165, 273, 295
uranium, 185ff
Uranus, vii, 27ff, 37, 50ff, 70, 145ff, 266, 272
   Centaurs and, vii, 28, 58, 146

V

Vallis Marineris, 35
Van Allen, 9, 50, 56, 66, 226, 305
Van Allen Belts, 9, 50, 66
van Noortwijk, 253
veal production, 208, 214
vegans, 214
vegetarianism, 214
Venera spacecraft, 33
Venus, vii, 26, 33ff, 70, 147ff, 224, 263ff, 298
   surface of, 33, 41
Venus Express spacecraft, 41
vermin of the skies, 46
vertebrates, 163
Virgo Cluster, 93
virtual particles, 128
viruses, 160, 218
vision, averted. *See also* naked eye observations, 23, 74, 165
volcanoes, 14, 35, 264
von Braun, M., 280
von Braun, W., 223, 226, 280
Voyager spacecraft, 53ff, 241
Vulcan, 28

W

walking, 213, 228
   upright, 165, 273, 295
Wallace, 110ff, 151, 153, 269, 305

warfare, 179, 186ff, 222, 233, 243ff, 304
   arms sales, 195
   chemical and biological, 189, 245
water, 11ff, 32ff, 56, 78, 112, 122ff, 138, 141, 149, 155ff, 176, 182, 200ff, 211, 216, 225ff, 252, 262ff, 287, 301
   drinking, 252
   on Earth, 20, 50, 176
water frost, 37
wealth, 202, 206, 215, 244, 245
   distribution of, 5, 151, 188, 191, 202, 214, 245ff
web sites, vi, 261, 265ff
Webb, 94
Webb Space Telescope, 94
whales, 203, 277
white dwarf, 76, 78, 93, 113, 135, 174, 176, 267, 305
Wigand, 211, 278
Wilkinson, 120, 122
Wilkinson Microwave Anisotropy Probe (WMAP), 120, 270
Wilson, i, 276
wind energy, 216
WMAP. *See* Wilkinson Microwave Anisotropy Probe
women, trafficking in, 159, 206, 213, 227, 280
Wong, K.
   Stranger in a New Land, 262, 273
World Health Assembly, 211, 245
World Trade Organization (WTO), 251
world view, 305
World War I, 186ff, 221, 224, 246
World War II, 186ff, 221, 224, 246
   freedom fighters, 251

X

X-rays, 89, 95, 294, 305

Y

yoga, 210, 213
Yucatan Peninsula, 180

Chicxulub, 180
Yucca Mountain, 275

Z

zero-degree-Kelvin radiation, 119
zero-G environment, 228

zero-point energy, 117, 216
zodiacal light, 23, 27

Made in the USA